Climate Change in Developing Countries

Results from the Netherlands Climate Change Studies Assistance Programme

Climate Change in Developing Countries

Results from the Netherlands Climate Change Studies
Assistance Programme

Edited by

M.A. van Drunen, R. Lasage and C. Dorland

*Institute for Environmental Studies
Vrije Universiteit
Amsterdam
The Netherlands*

www.cabi.org

CABI is a trading name of CAB International

Head Office
Nosworthy Way
Wallingford
Oxfordshire OX10 8DE
UK

Tel: +44 (0)1491 832111
Fax: +44 (0)1491 833508
E-mail: cabi@cabi.org
Website: www.cabi.org

North American Office
875 Massachusetts Avenue
7th Floor
Cambridge, MA 02139
USA

Tel: +1 617 395 4056
Fax: +1 617 354 6875
E-mail: cabi-nao@cabi.org

© CAB International 2006. All rights reserved. No part of this publication may be reproduced in any form or by any means, electronically, mechanically, by photocopying, recording or otherwise, without the prior permission of the copyright owners.

A catalogue record for this book is available from the British Library, London, UK.
A catalogue record for this book is available from the Library of Congress, Washington, DC.

Library of Congress Cataloging-in-Publication Data

Climate change in developing countries : results from the Netherlands
   Climate Change Studies Assistance Programme / edited by M.A.
   van Drunen, R. Lasage and C. Dorland
p. cm.
   Includes bibliographical references and index.
   ISBN-13: 978-1-84593-077-6 (alk. paper)
   ISBN-10: 1-84593-077-0 (alk. paper)
   1. Climate change—Developing countries. 2. Climate change
 • Government policy—Developing countries. I. Drunen, M. A. van.
   II. Lasage, R. III. Dorland, C. IV. Netherlands Climate Change
   Studies Assistance Programme.
   QC981.8.C5C511413 2006
   551.69172′4--dc22
2005031191

ISBN-10:   1-84593-077-0
ISBN-13: 978-1-84593-077-6

Typeset by MRM Graphics Ltd, Winslow, Bucks
Printed and bound in the UK by Biddles Ltd, King's Lynn

Contents

Preface — x

1. Introduction and NCCSAP Methodology — 1
    1.1 Introduction — 1
        1.1.1 Content — 2
        1.1.2 Authors — 2
    1.2 Approach of NCCSAP Phase 1 — 2
        1.2.1 Objectives — 2
        1.2.2 Responsibilities — 3
        1.2.3 The studies — 3
    1.3 Methodology for Emission Inventories — 5
        1.3.1 Policy background — 5
        1.3.2 Approach — 5
        1.3.3 Emission inventories in the NCCSAP — 6
    1.4 Methodology for Mitigation Assessment in the Energy Sector — 7
        1.4.1 Policy background — 7
        1.4.2 Approach in the NCCSAP — 8
    1.5 Adaptation Assessments — 11
        1.5.1 Policy background — 11
        1.5.2 The IPCC Common Methodology — 11
        1.5.3 The IVM/UNEP handbook — 12
        1.5.4 Adaptation assessment approaches in the NCCSAP — 12

2. Country Experiences and Highlights — 17
    2.1 Introduction — 17
    2.2 Bolivia — 18
        2.2.1 Introduction — 18
        2.2.2 Approach — 18
        2.2.3 Results — 19
        2.2.4 Experiences and lessons learned — 22
        2.2.5 Follow-up research — 23
        2.2.6 Policy implications — 24
        2.2.7 Conclusions — 25
    2.3 Colombia — 26
        2.3.1 Introduction — 26
        2.3.2 Approach — 26

2.3.3 Results of the Gulf of Morrosquillo study	27
2.3.4 Experiences and lessons learned	29
2.3.5 Follow-up research	30
2.3.6 Policy implications	31
2.3.7 Conclusions	32
2.4 Ecuador	34
2.4.1 Introduction	34
2.4.2 Approach	35
2.4.3 Results of the vulnerability assessment	35
2.4.4 Experiences and lessons learned	39
2.4.5 Follow-up research	40
2.4.6 Policy implications	40
2.4.7 Conclusions	41
2.5 Egypt	42
2.5.1 Introduction	42
2.5.2 Approach	43
2.5.3 Results of the vulnerability and adaptation assessment	44
2.5.4 Adaptation	45
2.5.5 Experiences and lessons learned	45
2.5.6 Follow-up research	46
2.5.7 Policy implications	47
2.5.8 Conclusions	47
2.6 Ghana	48
2.6.1 Introduction	48
2.6.2 Approach	49
2.6.3 Results of the water resources vulnerability study	51
2.6.4 Experiences and lessons learned	55
2.6.5 Follow-up research	55
2.6.6 Policy implications	55
2.6.7 Conclusions	57
2.7 Kazakhstan	58
2.7.1 Introduction	58
2.7.2 Approach	59
2.7.3 Results of the vulnerability assessment	59
2.7.4 Experiences and lessons learned	62
2.7.5 Follow-up research	63
2.7.6 Policy implications	63
2.7.7 Conclusions	63
2.8 Mali	64
2.8.1 Introduction	64
2.8.2 Approach	65
2.8.3 Results	66
2.8.4 Experiences and lessons learned	71
2.8.5 Follow-up research	71
2.8.6 Policy implications	71
2.8.7 Conclusions	72
2.9 Mongolia	73
2.9.1 Introduction	73
2.9.2 Approach	74
2.9.3 Results of the water resources assessment	74
2.9.4 Experiences and lessons learned	78
2.9.5 Follow-up research	79

2.9.6 Policy implications	80
2.9.7 Conclusions	81
2.10 Senegal	82
2.10.1 Introduction	82
2.10.2 Approach	83
2.10.3 Results of the coastal zone study	83
2.10.4 Experiences and lessons learned	86
2.10.5 Follow-up research	87
2.10.6 Policy implications	87
2.10.7 Conclusions	87
2.11 Suriname	88
2.11.1 Introduction	88
2.11.2 Approach	88
2.11.3 Results of the coastal zone management study	89
2.11.4 Experiences and lessons learned	92
2.11.5 Follow-up research	95
2.11.6 Policy implications	95
2.11.7 Conclusions	96
2.12 Vietnam	97
2.12.1 Introduction	97
2.12.2 Approach	99
2.12.3 Results of the coastal zone study	99
2.12.4 Experiences and lessons learned	103
2.12.5 Follow-up research	103
2.12.6 Policy implications	104
2.12.7 Conclusions	104
2.13 Yemen	106
2.13.1 Introduction	106
2.13.2 Approach	106
2.13.3 Results of climate change on agriculture in Yemen	107
2.13.4 Experiences, lessons learned and follow-up research	109
2.13.5 Policy implications	109
2.14 Zimbabwe	110
2.14.1 Introduction	110
2.14.2 Approach	111
2.14.3 Results of the Bindura Smelter and Refinery study	111
2.14.4 Experiences and lessons learned	113
2.14.5 Follow-up research	114
2.14.6 Policy implications	115
2.14.7 Conclusions	115
3. Cross-country Syntheses	**116**
3.1 Introduction	116
3.2 Emission Inventories	116
3.2.1 Overview	116
3.2.2 Bolivia	117
3.2.3 Kazakhstan	117
3.2.4 Suriname	118
3.2.5 Discussion and recommendations	118
3.3 Mitigation Assessment of the Energy Sector	119
3.3.1 Introduction	119
3.3.2 Analysis	119

3.3.3 Follow-up activities	122
3.4 Adaptation and Water Resources	123
3.4.1 Introduction	123
3.4.2 Adaptation strategies	126
3.5 Adaptation in Coastal Zones	127
3.5.1 Introduction	127
3.5.2 The impacts of SLR	128
3.5.3 Response strategies in general	129
3.5.4 Action plans defined in the vulnerability assessment	132
3.5.5 Feasibility of implementation of action plans	134
3.6 Adaptation and Land Use	135
3.6.1 Agriculture	135
3.6.2 Forestry	138
3.7 National Communications	140
3.7.1 Introduction	140
3.7.2 Bolivia	141
3.7.3 Costa Rica	142
3.7.4 Ghana	142
3.7.5 Mongolia	143
3.7.6 Senegal	144
3.7.7 Yemen	144
3.7.8 Synthesis	145

4. Evaluation, Lessons Learned and Outlook — 147

4.1 Introduction	147
4.2 Mitigation Assessment	147
4.2.1 Introduction	147
4.2.2 Recent developments with regard to the Clean Development Mechanism (CDM)	149
4.2.3 Future developments with regard to CDM	150
4.2.4 Long-term perspectives on mitigation activities	151
4.3 Adaptation Assessments	153
4.3.1 Coastal zones	153
4.3.2 Water resources	157
4.33 Agriculture and forestry	157
4.4 National Communications	158
4.4.1 Guidelines	158
4.4.2 The UNDP Adaptation Policy Framework (APF)	159
4.4.3 The National Adaptation Programmes of Action (NAPA) guidelines	159
4.4.4 Discussion	160
4.5 Capacity Building and Awareness Raising	161
4.5.1 Introduction	161
4.5.2 Concluding remarks on the questionnaires	161
4.6 NCCSAP in Comparison with Other Country Study Programmes	162
4.6.1 Introduction	162
4.6.2 US Country Studies Program	162
4.6.3 UNEP Country Studies on Climate Change Impacts and Adaptation Assessments	163
4.6.4 Comparing the programmes	165
4.7 Recommendations	166
4.7.1 Objectives	166
4.7.2 Involvement	167

    4.7.3 Livelihoods   168
    4.7.4 A suitable framework   168
    4.7.5 Conclusions   169

Appendix: Questionnaire Results   **170**
    Policymakers and Coordinators   170
    Researchers   174

References   **182**

Index   **187**

Preface

The Netherlands recognizes the need to assist developing countries, not only to comply with its own obligations under the United Nations Framework Convention on Climate Change (UNFCCC) but also to enable these countries to develop effective strategies to mitigate climate change and adapt to the adverse effects thereof. Therefore, the Netherlands Climate Change Studies Assistance Programme (NCCSAP) was launched in 1996. The first phase of the NCCSAP is almost finished and the second phase, which has slightly different objectives, has already started.

The Institute for Environmental Studies (IVM), which manages NCCSAP-I, took the initiative to compile this book. An overview is presented of the most interesting results of the climate studies carried out in the countries involved. Furthermore a critical evaluation is given of the methodologies and approaches, and recommendations for future studies about climate change in developing countries. This book does not pretend to be a summary of all NCCSAP studies; rather, it gives the reader an impression of the work that has been carried out in the last 8 years within the programme.

This book is meant for everyone who is involved in climate-related projects in developing countries. Both researchers and policymakers can benefit from the experiences and evaluations of the NCCSAP studies. Subjects dealt with include not only impact studies but also vulnerability and adaptation, mitigation and climate-related policy. More about NCCSAP, in particular about the second phase, can be found at www.nlcap.org

In line with the objectives of the NCCSAP, the core part of this book was written by the project members in the developing countries themselves. The editors thank all the authors, several of whom had quite different jobs by the time they were writing their contribution. The technical consultants also played an essential role, especially Jan Verhagen, Nico van der Linden, Marcel Rozemeijer and Arjan van der Weck; they put tremendous efforts into this book. Ann Holleman is greatly acknowledged for correcting the 'Dunglish' and other grammatical errors, and Laurens Bouwer is thanked for reviewing the report internally.

The NCCSAP would not exist without the late Jan Feenstra. He not only invented the impossible acronym NCCSAP, but he was also involved in the methodology development for climate impact and adaptation studies, and for the organizational structure of the programme. In the mid-1990s he managed the programme and travelled the world to make sure that all studies went well.

Finally, the editors thank the Netherlands Ministry of Foreign Affairs, Directorate-General for International Cooperation (DGIS), which recognized the importance of climate change for developing countries and decided to initiate the NCCSAP. One of the NCCSAP products is this book.

Michiel van Drunen, Manager NCCSAP-I
September 2005

1
Introduction and NCCSAP Methodology

1.1 Introduction

The Netherlands recognizes the need to assist developing countries and countries with economies in transition, not only to comply with its own obligations under the United Nations Framework Convention on Climate Change (UNFCCC) but also to enable these countries to develop effective strategies to mitigate climate change and adapt to the adverse effects thereof. The Netherlands Climate Change Studies Assistance Programme (NCCSAP) was launched in 1996. This programme aims to help developing countries to prepare their National Communications and to undertake capacity building, education and training activities (VROM, 2001).

NCCSAP was preceded by four coastal zone management studies in Bangladesh, Egypt, Nicaragua and Vietnam. These studies focused on the effects of sea level rise (SLR) in coastal areas, possible adaptation options and capacity building. The Coastal Zone Management Centre in The Hague managed these studies. In this book the main results of the studies in Egypt and Vietnam are summarized in Chapter 2.

NCCSAP includes two phases and in this book only the first phase is described. In total 13 countries participated in the programme and at the time of writing this book (summer 2004) only the studies in Bhutan had not yet been completed. The second phase started in 2003 with a slightly different focus and partly different countries.

The NCCSAP-I studies resulted in many reports considering climate change issues for each study in each country. Existing information was summarized, new information was generated and information gaps were identified. Among these reports are National Communications, National Action Plans and emission inventories, mitigation studies and sectoral impact and adaptation studies. The sectoral reports and the policy documents were generally of high quality. The policy documents were important to fulfil the commitments to the UNFCCC and to include climate change issues into existing plans for sustainable development in the countries examined.

This book provides a comprehensive information source on NCCSAP and describes the achievements and experiences in the countries involved. We hope it also serves as an invaluable source of information to climate change experts and policymakers, as well as the Directorate-General for International Cooperation (DGIS) and other donors active in the context of the UNFCCC. It presents summaries of the studies as well as the experiences and lessons learned from the NCCSAP. Thus far, many detailed reports have been prepared in the national language of the countries participating and only a few scientists have published their work in international journals. Therefore, this book also aims to provide valuable input to the Intergovernmental Panel on

Climate Change (IPCC) Fourth Assessment Report.

The book covers emission inventories, mitigation and adaptation, but the emphasis is on adaptation. This is in line with the recent developments in the UNFCCC, where adaptation, sustainable development and the links between adaptation and poverty alleviation are getting higher priority and where new funds for adaptation are created. Additionally new methods and guidelines for adaptation assessment with a focus on policy development, livelihood systems and poverty are currently being developed and pilot studies and adaptation projects are being started, e.g. in NCCSAP-II. Here, we present the state of the assessment methodologies and policies in the 1990s and describe possible next steps. The next section provides an overview of the book contents.

1.1.1 Content

This book includes the following sections:

1. Introduction and methodology
2. Country experiences and highlights
3. Cross-country syntheses
4. Evaluation, lessons learned and outlook

The remainder of Chapter 1 provides a general introduction to the NCCSAP. It also provides summaries of the methodologies used in the studies.

Chapter 2 is the core of the book. Here, representatives of the countries have summarized one of their studies and reflected on the strengths and weaknesses of the approach they adopted.

Chapter 3 presents a cross-country synthesis for each of the sectors assessed in the countries. It also provides a cross-country analysis of the National Communications.

Chapter 4 evaluates the activities in the programme in terms of content and process. Topics addressed include capacity building in the participating countries, coordination and programme management, scientific quality of studies, policy relevance of activities, contribution to climate policy development, contribution to awareness raising, donor activities and policy, etc. Experiences and lessons learned are identified and discussed. Furthermore, recommendations for future activities are discussed.

1.1.2 Authors

The authors of the book are the researchers and policymakers who were actually involved in the studies. The core of this book, Chapter 2, was written by the project manager or other senior researchers of the country concerned. Chapters 1, 2 and 4 were written by the consultants who assisted the local researchers with training, visits and backstopping. The editors were involved in the NCCSAP project management. Section authors are indicated for each section unless the section was written by an editorial team.

1.2 Approach of NCCSAP Phase 1

1.2.1 Objectives

The NCCSAP is an initiative of the Netherlands Ministry of Foreign Affairs, Directorate-General for International Co-operation (DGIS), and started in 1996. The objectives of the first phase of the programme (van Drunen and Dorland, 2000) were:

- to enable developing countries to implement commitments under the Framework Convention on Climate Change;
- to create a greater awareness of climate change issues; and
- to increase the involvement of policymakers, scientists and the general public.

The programme supported the responsible ministry, mostly the Ministry of Environment, to initiate climate change studies that were carried out by appropriate scientific institutions. In 2003 the second phase, which is not treated here, started with different objectives (ETC International, 2004).

Based on the conclusion of the IPCC (McCarthy et al., 2001), it is expected that the impacts of climate change will be most severe in developing countries and may thereby hamper sustainable development and lead to acute food shortage, poverty and health hazards. The NCCSAP provided opportunities to carry out in-depth climate change impact and adaptation studies. The scope of the studies depended on national needs, priorities, experiences and expertise.

The objective was that the results of the climate change studies would find their way into the national sustainable development plans, environmental action plans and the National Communications to the UNFCCC of the countries participating. The studies also assisted the national institutes and authorities in strengthening their institutional roles and responsibilities by improving the public awareness on climate change issues.

A prerequisite was that the studies were carried out by institutions and scientists of the participating countries themselves to ensure capacity building and ownership. The national study teams were assisted by international consultants for each specific sector of the climate change studies.

1.2.2 Responsibilities

The Institute for Environmental Studies (IVM) was contracted by DGIS to manage the first phase of the NCCSAP in cooperation with the Netherlands Coastal Zone Management Centre (CZMC). The following countries participated: Bolivia, Bhutan, Colombia, Costa Rica, Ecuador, Ghana, Kazakhstan, Mali, Mongolia, Senegal, Suriname, Yemen and Zimbabwe. The Ministry of Foreign Affairs (DGIS) selected the countries.

The managing institute (IVM or CZMC) contracted with the responsible ministry. The local ministry appointed a *national focal-point coordinator* who was responsible for the management of all studies in a country. This coordinator was often involved in the studies, but he or she also subcontracted to other institutes in the country. The focal-point coordinator frequently reported the progress and expenditures of the studies.

The studies in each country started with a mission: experts from IVM and/or CZMC visited the country to assist with the preparation of the project proposal or work plan. The philosophy behind these missions was to give the projects a good start: to clarify what needed to be done and who was responsible for the activities agreed upon.

After formal inclusion in the NCCSAP by the Ministry of Foreign Affairs, the national focal point elaborated a detailed plan of operations, in close cooperation with the management of the programme and the international consultants. This contained the organizational framework of the country study, project descriptions, terms of references, work plans and budgets.

1.2.3 The studies

In each country, the studies were carried out by local specialists who were assisted by international consultants. The consultants actively supported the local researchers through on-site visits, organized training sessions and seminars. They assisted the studies through e-mail contact and reviewed draft documents. In some cases local experts were sent to specific training sessions or workshops. Table 1.1 gives an overview of the studies per country.

Many of the studies were closely related. For instance, before effective mitigation measures can be formulated, an emission inventory must be carried out. Also, water resources data are needed to formulate adaptation options in agriculture, economic and climate scenarios. This is shown schematically in Fig. 1.1.

Because IVM was closely involved in the handbooks on inventories (Houghton et al., 1997) and on vulnerability and adaptation (Feenstra et al., 1998), and because of its experience in former studies, it was able to provide the focal points with technical, practical and scientific suggestions and methodologies.

Besides the technical, sectoral studies, the country studies consisted of two or three national workshops where scientists, policymakers, and representatives of NGOs and the press met to exchange information regarding the set up of the studies and the results, and to discuss the implications for national policy. In this way, not only policymakers and scientists were informed about climate change issues, but also the general public and other stakeholders.

Table 1.1. Overview of studies in the NCCSAP.

Country	GHG emissions		Vulnerability and adaptation				Policy
	Emission inventory	Mitigation	Agriculture	Forestry natural areas	Coastal zones	Water resources	National Communication
Bhutan			x			x	
Bolivia	x	x	x	x		x	x
Colombia					x		
Costa Rica			x	x	x		x
Ecuador					x	x	
Ghana					x	x	x
Kazakhstan	x			x[a]	x		
Mali			x			x	
Mongolia		x	x			x[b]	x
Senegal			x		x		x
Suriname	x				x	x	
Yemen		x	x			x	x
Zimbabwe		x	x				

[a]The study concerned mudflows and avalanches.
[b]The water resources study was not formally included.

In addition to the national workshops, the programme provided regional workshops (e.g. the workshop in Latin America with Bolivia, Costa Rica, Ecuador and Suriname) and bilateral meetings between countries. IVM, the US Country Studies Program, the United Nations Development Programme (UNDP), the United Nations Environment Programme (UNEP) and the Japanese government organized the international workshop 'National assessment results of climate change: impacts and responses' in Costa Rica from 25–28 March, 1998. Representatives from most countries participated in this workshop.

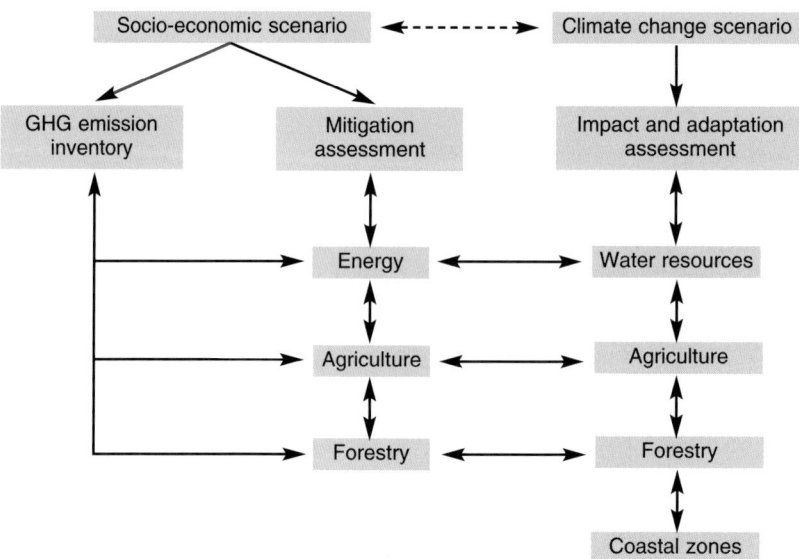

Fig. 1.1. Interdependence of climate change study results.

In the following sections we describe the methodologies followed in the sectors indicated in Table 1.1.

1.3 Methodology for Emission Inventories

Michiel van Drunen[1]

1.3.1 Policy background

Greenhouse gas (GHG) emission inventories play a key role in the UNFCCC. Since the UNFCCC aims to stabilize GHG concentrations in the atmosphere at a level that would prevent dangerous anthropogenic (human-induced) interference with the climate system, information is required about emissions of GHGs and also the trends in these emissions. Therefore,

> Each non-Annex I Party shall ... communicate to the Conference of the Parties a national inventory of anthropogenic emissions by sources and removals by sinks of all greenhouse gases (GHGs) not controlled by the Montreal Protocol, to the extent its capacities permit ... Non-Annex I Parties should use the *Revised 1996 Intergovernmental Panel on Climate Change (IPCC) Guidelines for National Greenhouse Gas Inventories*, hereinafter referred to as the IPCC Guidelines, for estimating and reporting their national GHG inventories.
>
> (UNFCCC, 2004a)

Hence, where Annex I countries (the developed countries) must provide annual reports, non-Annex I countries must report on their emissions 'to the extent their capacities permit'. Emission data are published on the website of the UNFCCC (unfccc.int). All countries in the NCCSAP are non-Annex I countries.

1.3.2 Approach

Emissions inventories were carried out in Bolivia, Kazakhstan and Suriname. As obliged by the UNFCCC, all used the revised 1996 IPCC guidelines (Houghton et al., 1997). The IPCC guidelines were first accepted in 1994 and published in 1995. The UNFCCC Third Conference of Parties (COP3), held in 1997 in Kyoto, reaffirmed that the *Revised 1996 IPCC Guidelines for National Greenhouse Gas Inventories* should be used as 'methodologies for estimating anthropogenic emissions by sources and removals by sinks of greenhouse gases' in calculation of legally binding targets during the first commitment period.

The gases covered in the Guidelines are the direct GHGs, carbon dioxide (CO_2), methane (CH_4) and nitrous oxide (N_2O), the indirect GHGs, carbon monoxide (CO), oxides of nitrogen (NO_x), non-methane volatile organic compounds (NMVOCs), halocarbons (HFCs, PFCs), sulphur hexafluoride (SF_6) and sulphur dioxide (SO_2). Halogenated species (i.e. chlorofluorocarbons (CFCs), hydro-chlorofluorocarbon 22 (HCFC-22), the halons, methyl chloroform and carbon tetrachloride) are not included because of parallel reporting requirements of countries in compliance with commitments under the Montreal Protocol.

The calculation of CO_2 emissions from fuel combustion may be done at three different levels, referred to as Tiers 1, 2 and 3 in the IPCC guidelines. Tier 1 methods, as used in the studies described here, concentrate on estimating the emissions from the carbon content of fuels supplied to the country as a whole (the Reference Approach, sometimes referred to as 'top-down') or to the main fuel combustion activities (called Emissions by Source Categories or 'bottom-up').

The Reference Approach

Carbon dioxide emissions are produced when carbon-based fuels are incinerated. National emissions estimates are based on amounts of fuels used and the carbon content of fuels. Hence the total fuel supplies to a country (imports and extraction) provide an accurate estimation of national CO_2 emissions.

Accounting for carbon is based mainly on the supply of primary fuels (i.e. fuels which are found in nature such as coal, crude oil, natural

---
[1] Institute for Environmental Studies, Vrije Universiteit, De Boelelaan 1087, 1081 HV Amsterdam, The Netherlands. Tel.: +31-20-5989 534, Fax: +31-20-5989 553, Email: michiel.van.drunen@ivm.falw.vu.nl

gas) and the net quantities of secondary fuels (e.g. gasoline and lubricants) brought into the country. To calculate the supply of fuels to the country the following data for each fuel and year chosen are required (Houghton et al., 1997):

- the amounts of primary fuels produced;
- the amounts of primary and secondary fuels imported;
- the amounts of primary and secondary fuels exported;
- the amounts of fuel used for international marine and aviation bunkers; and
- the net increases or decreases in fuel stocks.

Emissions by Source Categories

A sectoral breakdown of national CO_2 emissions using the defined IPCC source categories is needed for monitoring and abatement policy discussions. The IPCC Reference Approach provides a quick estimate of the total CO_2 emissions from fuels supplied to the country but it does not break down the emissions by sector. The development of a Tier 1 method giving non-CO_2 emissions from fuel combustion by sector has been extended to CO_2 so that sectoral information can be obtained simply for this gas (Houghton et al., 1997).

The following sectors are considered: energy, industrial processes, solvents and other product use, agriculture, land use change and forestry, and waste. Crucial in the emission estimations are *emission factors*, which link the activity data (e.g. production data in tonnes or kWh) to the GHG emissions. Although the IPCC provides default emission factors, these tend to vary from country to country because they depend heavily on the technologies used and the specific conditions under which they are operated.

Reporting

Documentation standards are necessary to ensure transparency of national inventories and to allow inventories to be reviewed. Therefore, one of the three Guideline workbooks is completely devoted to reporting instructions (Houghton et al., 1997). The most important output tables include the GHG emissions by sector and the total emissions by gas type (CO_2, CH_4, etc.). In several output tables the emissions of the non-CO_2 gases are expressed in CO_2 equivalents (CO_2e), i.e. the emissions are multiplied by the global warming potential (GWP) of the GHG considered. By definition the GWP of CO_2 is 1. The GWP of CH_4 and N_2O are 21 and 310, respectively (Houghton et al., 1997).

1.3.3 Emission inventories in the NCCSAP

Dorland et al. (2001) present an overview of the approach and the results of the inventories. Below, we have summarized the methodological approaches in the countries where emission inventories were carried out within the NCCSAP framework.

Bolivia

The Bolivian inventory of GHG emissions was conducted in accordance with the 1996 revised IPCC guidelines. Unlike the other inventories, Module 3, Use of Solvents and Other Products, has been included.

For most but not all sectors, IPCC default emission factors were applied where necessary. For deforestation a more suitable, local value has been used. The report has a specific section on uncertainty, in which the uncertainty in the CO_2 emission figure for deforestation is estimated to be 35%. In view of the huge contribution of this particular emission source, further research is needed to get a clearer view of the exact magnitude of deforestation-related CO_2 emissions in Bolivia.

Kazakhstan

The revised IPCC guidelines of 1996 were used as the methodological basis. In some cases the IPCC methodology was complemented to reflect national circumstances and/or data availability. The GHG emission inventory is divided into six categories, but solvent and other product use was not included because of a lack of available data.

Conversion factors to calculate CO_2 emissions were taken mostly from local literature sources. The main limitations in the studied inventory, however, are the lack of verified local emission factors, particularly for non-CO_2 emissions. Another limitation of the study results from activities for which neither a local nor an IPCC emission factor exists. Finally, lack of data and high aggregation levels for certain data form another limitation of the study.

Suriname

IPCC default values for fuel type and combustion efficiency were used for the calculation of CO_2 emissions. Solvent and other product use was omitted because of lack of available data. Many other data gaps were identified. The only industry that was reasonably accurately described was the aluminium industry. Only default values were available for the agricultural and the waste sectors and detailed information on land use was not available.

In the report, no attempt was made to compare CO_2 to other GHG emissions. Neither was an attempt made to address uncertainties, but considering the data gaps mentioned and the frequent use of non-area specific default values, the uncertainty forces one to be very careful in drawing any definite conclusions from the figures presented.

1.4 Methodology for Mitigation Assessment in the Energy Sector

Nico van der Linden[2] and Jan-Willem Martens[3]

1.4.1 Policy background

The UNFCCC (UN, 1992) does not oblige non-Annex I countries (developing countries) to mitigate their GHG emissions. However, Article 4.5 states:

> The developed country Parties and other developed Parties included in Annex II shall take all practicable steps to promote, facilitate and finance, as appropriate, the transfer of, or access to, environmentally sound technologies and know-how to other Parties, particularly developing country Parties, to enable them to implement the provisions of the Convention. In this process, the developed country Parties shall support the development and enhancement of endogenous capacities and technologies of developing country Parties. Other Parties and organizations in a position to do so may also assist in facilitating the transfer of such technologies.

The Clean Development Mechanism (CDM) was established under Article 12 of the Kyoto Protocol. It is the only flexible mechanism in the Kyoto Protocol that involves non-Annex I countries. Activities that reduce emissions in non-Annex I countries can result in Certified Emission Reductions (CERs), which may be transferred to Annex I countries. For the countries (or businesses) buying the emission rights, using the CDM offers a clear advantage. When the price of the emission rights is lower than the costs associated with internal or on-site emission reductions, buying credits lowers the overall costs of compliance.

To the investor in CDM projects, the opportunity to sell credits offers an extra source of revenue above the normal project revenues, e.g., for selling electricity. This will increase the financial viability of the project. What is often very important for investments in developing countries is that part of the revenue from the project will come in hard currency and from a buyer with a good reputation (such as the World Bank, a European government or a private company). The prospect of this revenue source can help to promote other financing for the project. For most project types, such as renewable energy projects, the carbon revenues will cover only 5% to 15% of the investment costs. For CH_4 emission reduction projects, the carbon revenues can cover more than half the costs of the project. This difference in extra return can have an effect on the types of projects that are developed under the CDM. The CDM is already operational.

---

[2] ECN Policy Studies, Badhuisweg 3, 1031 CM Amsterdam, The Netherlands. Tel.: +31 (0)224 564431, Fax: +31 (0)20 4922812, Email: n.vanderlinden@ecn.nl
[3] EcoSecurities, Mauritskade 25, 2514 HD Den Haag, The Netherlands. Tel.: +31 (0)70 3654749, Fax: +31 (0)70 3656495, Email: nl@ecosecurities.com

Under guidance of the CDM Executive Board procedures have been developed and the first projects are ready for approval.

Perhaps the most important benefit provided by CDM for host countries is that they serve to attract foreign investments in low-emission technologies. Both the direct investments and the resulting improvements in efficiency have a positive contribution to the economy and can lead to more employment. Another benefit is that CDM projects often also reduce other pollution in the host country, such as local air pollution. For example, a waste management project reduces CH_4 emissions, but also reduces odours around the landfill and reduces groundwater pollution.

1.4.2 Approach in the NCCSAP

Introduction

A mitigation assessment involves an analysis of costs, benefits and reduction potential of options that can be used to reduce GHG emissions. These options may either reduce the amount of GHG emissions or increase the storage of carbon (sequestration). In the NCCSAP Phase I study only the former type of options was taken into account.

In principle, two approaches can be adopted for mitigation assessment: the top-down approach and the bottom-up approach. The top-down approach starts at the national level and takes into account the interactions between the energy sector and the other economic sectors. The approach involves an analysis of macro-variables such as sectoral growth rates and past improvements in energy efficiency to estimate the emission reduction potential. The bottom-up approach, on the other hand, focuses on individual technologies and attempts to estimate the costs and reduction potential of these technologies. This approach results in a detailed list of concrete reduction options, but no interactions with other economic sectors are taken into account. In the NCCSAP Phase I study the bottom-up approach has been adopted for mitigation assessment to assess in detail the impact on GHG emission reduction of individual technological options.

A mitigation assessment consists of the following key components:

1. Development of a national GHG emissions inventory.
2. Identification of suitable emission reduction options.
3. Assessment of costs and GHG reduction potential of options.
4. Design of aggregated cost abatement curve.
5. Barrier analysis.
6. Strategy formulation.

In the framework of the NCCSAP Phase I study, assistance in mitigation assessment has been provided to the mitigation teams of Bolivia, Yemen, Zimbabwe (see also Section 2.14) and Mongolia. In Bolivia and Yemen the focus of assistance was on the assessment of costs and GHG reduction potential of identified emission reduction options. This involved training on the Long Range Energy Alternative (LEAP) model and the development of a LEAP version that could be used to evaluate the identified emission reduction options. For Zimbabwe and Mongolia these activities were already completed in the framework of other programmes and the assistance focused on barrier analysis and strategy formulation.

Development of a national emission inventory

To get insight into the potential for mitigating GHGs and which costs are necessary to realize this potential, it is necessary to have a national emission inventory (see also Section 1.3). The emission inventory can be used to identify existing GHG emitting sources that could be replaced with technologies that use less energy per unit of output.

Identification of emission reduction options

Potential options to reduce GHG emissions can be divided into two broad categories: technological and non-technological options. The technological options can be further divided into options that save energy by improving the efficiency of the process and options aimed at a switch to a cleaner fuel. Non-technological options attempt to influence the level of individual energy consumption by providing financial incentives (tax, subsidy) or by enhancing awareness with regard to environmental issues. In this step, experts on the various economic sectors are asked to identify potential emission reduction options that will be included in the mitigation assessment.

Assessment of costs and GHG reduction potential

The criterion used to rank the identified options is the net costs (benefits minus costs) per tonne of CO_2 reduction, which is defined as the ratio of the total net costs of an option divided by the total emission reduction generated by the option.

Cost–benefit analysis is applied to estimate the net costs and involves a systematic comparison of all costs and benefits of a reduction option from a national point of view. Costs include investment costs and operation and maintenance costs of the emission reduction option. Benefits include the reduction in expenditure on fuels due to a more efficient technology and the benefits related to a reduction of GHGs. Because the latter type of benefits cannot be properly monetized, cost–benefit analysis for mitigation assessment is reduced to cost-effectiveness analysis, i.e. the analysis of achieving the stated objective in the least costly manner. The cost effectiveness is an important criterion in a comparison of the options and is determined by dividing the present value of the net costs by the emission reduction. For some options the net costs are negative (benefits exceed costs), indicating that the end user will financially benefit if the option is implemented. There could, however, be several reasons why these 'win-win' options are not taken autonomously, e.g. because they require large investments, or because there is a lack of knowledge about them.

The total emission reduction of options can be estimated by a comparison of baseline and mitigation scenarios. A baseline scenario describes the future situation in which no policies or programmes will be implemented to reduce GHG emissions. A baseline scenario is not simply an extrapolation of past trends; it also includes autonomous developments that will occur even without policy interventions. Because it is extremely difficult to predict future economic developments, alternative baseline scenarios usually reflect different assumptions on, among others, factors such as future economic growth (low, medium, high).

A mitigation scenario describes the future situation based on the assumption that policy options to reduce GHG emissions are implemented. These options may include technological options and non-technological options and several mitigation scenarios can be developed to assess the impact of individual options. LEAP was used as a tool to develop baseline and mitigation scenarios in Bolivia and Yemen. LEAP is a technology-based accounting model that enables the analyst to store the collected data in a structured and clear manner and to quickly quantify the effects of mitigation options.

Design of an aggregated cost abatement curve

The CO_2 abatement cost curve provides a ranking of individual reduction options based on their net cost effectiveness. This curve shows the relationship between net abatement costs per tonne of CO_2 and the total quantity of the emission reduction. In some cases, the net abatement cost curve starts at negative values for net costs per unit reflecting the fact that some of the options have marginal net benefits.

Figure 1.2 shows an example of an aggregated cost abatement curve. It depicts the projected CO_2 cost abatement curve for all non-Annex I countries for options in the unit cost range of –50 to +50 US$/t CO_2 equivalents. The curve is based on an inventory of options in 24 non-Annex I countries to reduce GHG emissions[4], and an extrapolation of the

---

[4] For a more detailed description of how the cost curve has been developed and an overview of the limitations of the curve see: ECN (1999) *Potential and Cost of Clean Development Mechanism Options in the Energy Sector: Inventory of Options in Non-Annex I Countries to Reduce GHG Emissions*; ECN-C-99-095.

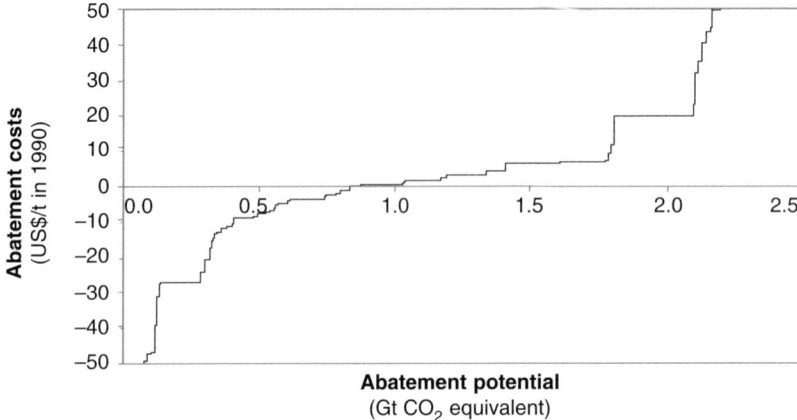

Fig. 1.2. The projected CO_2 abatement costs for all non-Annex I countries.

projected aggregated CO_2 abatement cost curve of the 24 countries to the remaining non-Annex I countries.

The total annual GHG emissions abatement potential in the non-Annex I countries in the first budget period (2008–2012) at unit costs up to US$50/t CO_2 has been projected at approximately 2.25 Gt CO_2 equivalents. Most of this potential is projected to be achievable at quite low costs. Up to 1.6 Gt/year appears feasible at costs of US$6/t or lower.

Barrier analysis

The mitigation analysis identifies and evaluates emission reduction options. However, existing barriers could prevent the implementation of the best options in terms of cost effectiveness. This obviously is the case for options with negative net costs. Implementation of these 'win-win' options can be justified on purely economic grounds. However, one should consider carefully *why* these options are not implemented and whether the estimate of the costs might neglect implementation barriers that can only be scaled at substantial additional costs. The extent to which these barriers can be removed effectively is of crucial importance to determine the real emission reduction potential and associated costs of identified options. Several types of barriers can be distinguished:

- financial barriers;
- policy and regulatory barriers;
- institutional barriers;
- technical barriers;
- lack of awareness.

The starting point for barrier analysis is an identification of existing barriers. This requires a detailed assessment of the reasons why potential attractive options are not implemented. Once the reasons are known, the challenge is to formulate policy measures that remove these barriers and create an enabling market environment for investments in emission reduction options.

Strategy formulation

Once all options have been ranked according to their cost effectiveness, a strategy can be formulated to achieve the agreed emission reduction target. Strategy formulation involves the design of a package of options that meets the required emission reduction. In addition to the cost effectiveness, several other aspects that are not captured by cost effectiveness (such as implementation barriers, required investments, social impact, additional tax burden) are taken into account in designing the most appropriate package. By assigning relative weights to the various aspects, options can be compared with each other.

1.5 Adaptation assessments

Jan Verhagen[5], Michiel van Drunen[6], Frits Mohren[7], Marcel Rozemeijer[8] and Arjan van der Weck[9]

1.5.1 Policy background

Article 4.4 of the UNFCCC states that 'The developed country Parties and other developed Parties included in Annex II shall also assist the developing country Parties that are particularly vulnerable to the adverse effects of climate change in meeting costs of adaptation to those adverse effects'. Article 4.8 adds the following:

> In the implementation of the commitments in this Article, the Parties shall give full consideration to what actions are necessary under the Convention, including actions related to funding, insurance and the transfer of technology, to meet the specific needs and concerns of developing country Parties arising from the adverse effects of climate change and/or the impact of the implementation of response measures, especially on:
>
> 1. Small island countries.
> 2. Countries with low-lying coastal areas.
> 3. Countries with arid and semi-arid areas, forested areas and areas liable to forest decay.
> 4. Countries with areas prone to natural disasters.
> 5. Countries with areas liable to drought and desertification.
> 6. Countries with areas of high urban atmospheric pollution.
> 7. Countries with areas with fragile ecosystems, including mountainous ecosystems.
> 8. Countries whose economies are highly dependent on income generated from the production, processing and export, and/or on consumption of fossil fuels and associated energy-intensive products.
> 9. Land-locked and transit countries.
>
> Further, the Conference of the Parties may take actions, as appropriate, with respect to this paragraph.

(UN, 1992)

All countries in the NCSSAP meet at least four of the criteria mentioned in Article 4.8.

1.5.2 The IPCC Common Methodology

The IPCC Common Methodology (the Seven Steps Method – IPCC, 1991, 1992a; Carter et al., 1994) offers a general approach to assess the impact of SLR. It can be applied to different countries (especially developing countries) or different areas in a country to compare different vulnerability profiles or to compose overall vulnerability profiles. It can also be used to assess vulnerability and adaptation in different levels of detail, depending on, among others, data availability for quantitative and qualitative assessments. Moreover, the method can be tailor-made to the specific circumstances and needs of a country. The Seven Steps Method includes the following steps:

1. The delineation of the case study area in the country and the specification of the SLR and the climate change boundary conditions.
2. The inventory of the study area characteristics yielding both the natural system data and the socio-economic data.
3. The definition of the relevant development factors and the economic scenarios of development.
4. The assessment of physical changes, socio-economic impact and natural system responses.
5. The formulation of response strategies and action plans.
6. The implementation feasibility of the

---

[5] Plant Research International, PO Box 16, 6700 AA Wageningen, The Netherlands. Tel.: +31 317 477001, Fax: +31 317 418094, Email: adrianus.verhagen@wur.nl

[6] See Note 1.

[7] Forest Ecology and Forest Management Group, Dept. of Environmental Sciences, Wageningen University and Research Centre, PO Box 342, NL-6700 AH Wageningen, The Netherlands. Tel.: +31 317 478026, Fax: +31 317 478078, Email: frits.mohren@btbo.bosb.wau.nl

[8] Water4ruimte consultancy, Ruychaverstraat 36[rood], 2013 GG Haarlem, The Netherlands. Tel.: +31 (0)23-5313547, Email: marcel.rozemeijer@water4ruimte.nl

[9] WL Delft Hydraulics, PO Box 177, 2600 MH Delft, The Netherlands. Tel.: +31 (0)15 2858585, Fax: +31 (0)15 2858582, Email: arjan.vdweck@wldelft.nl

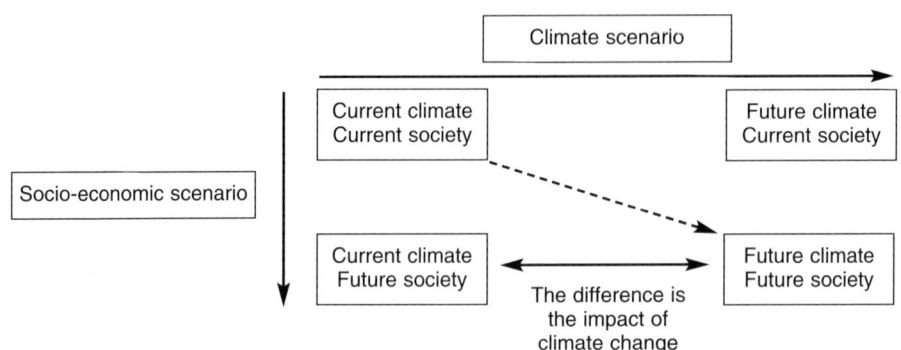

Fig. 1.3. Climate and socio-economic scenarios (Source: Feenstra *et al.*, 1998).

action plans. Here several structural aspects of the society of a country are evaluated to determine whether the action plans can be implemented or whether the implementation will encounter difficulties.

7. The identification of the types of assistance needed based on the potential problems encountered in the implementation of the action plans.

1.5.3 The IVM/UNEP handbook

The IVM/UNEP handbook *Handbook on Methods for Climate Change Impact Assessment and Adaptation Strategies* (Feenstra *et al.*, 1998) was the first comprehensive guide for adaptation assessments in multiple sectors. It sets out a methodology based on climate scenarios and socio-economic scenarios (Fig. 1.3). It includes separate chapters on water resources, coastal zones, agriculture, rangeland and livestock, human health, energy, forest, biodiversity and fisheries. The coastal zone section is 'based on a combination of widespread experience using the Common Methodology and other methods for coastal vulnerability assessment, which have been developed in response or addition to the Common Methodology' (Subsection 1.5.2).

1.5.4 Adaptation assessment approaches in the NCCSAP

The adaptation assessments of the coastal zone studies in the pre-NCCSAP and the NCCSAP studies closely followed the IPCC Common Methodology (Subsection 1.5.2). In Chapter 2, six coastal zone studies are described: in Colombia (Section 2.3), Ecuador (Section 2.4), Egypt (Section 2.5), Senegal (Section 2.10), Suriname (Section 2.11) and Vietnam (Section 2.12). Section 3.5 provides a synthesis of the coastal zone studies. In the following subsections the approach of the water resources, agriculture and forestry projects will be presented.

Water resources

In the water resources sector, technology, economics and institutions interact to make water supply meet water demand. In managing water resource systems, water managers ask, 'Can we modify the management of current systems to adapt to climate change?' and 'How might climate change impact the design of new water resource infrastructure?'. The water resources sector by its nature is very adaptive, on various time and spatial scales. Also, water managers have a wealth of knowledge and experience managing under changing hydrologic and socio-economic conditions. This experience places them in a good position to adapt the operation of their systems to a change in climate, if that change is not too great or too rapid (Feenstra *et al.*, 1998).

BIOPHYSICAL IMPACTS. The main components of the hydrologic cycle are precipitation, evaporation and transpiration. Changes in the climate parameters such as solar radiation, wind, temperature, humidity and cloudiness will affect evaporation and transpiration. Changes in evapotranspiration and precipitation will affect the amount and the distribution, spatially and temporally, of surface runoff.

Climate change can affect water quality in three ways. First, reduced hydrologic resources may leave less dilution flow in the stream, leading to degraded water quality or increased investments in wastewater treatment. Second, higher temperatures reduce the dissolved oxygen content in water bodies. Third, in response to climate change, water uses, especially those for agriculture, may increase the concentration of pollution being released to the streams. Together, these pose a threat to water quality and the integrity of the aquatic ecosystem.

SOCIO-ECONOMIC IMPACTS. Water use is generally divided into non-market and market uses. Non-market water uses are aesthetic uses, certain recreational uses and aquatic ecosystem integrity. Market water uses can be aggregated into five major water use sectors:

1. Agriculture: irrigation and livestock.
2. Industry: industry, mining, navigation, recreation.
3. Energy: thermoelectric cooling and hydroelectric power.
4. Municipal: public supply, domestic and commercial.
5. Reservoir.

An additional market use is dilution water for pollution abatement. It is typically considered a market use because it can be valued at the cost savings of additional waste treatment to meet water quality standards.

The water management system (i.e. water supply system) is made of two parts: surface water and groundwater. Although they are linked at the river basin water balance level, they are distinct in the water supply infrastructure. Climate change can affect the surface water supply via reduced flows into the storage reservoir or increased variability in inflow, which will affect firm yields from existing storage facilities. An additional impact in arid and semi-arid regions could be increased reservoir evaporative losses. The groundwater supply will be affected by increased or decreased percolation of water due to changes in the amount and distribution of precipitation and stream flow. This can lead to increased pumping costs if percolation decreases because of decreased precipitation or soil moisture loss.

With great uncertainties about the local and regional impacts of climate change on hydrologic resources and uncertain future water demands driven by socio-economic change, an assessment of climate change impacts on water resources is a complex process. In addressing the sensitivity of water resources to changes in climate, the biophysical and socio-economic conditions must be considered (Feenstra et al., 1998).

The goal and scope definition includes the following steps:

- Select the exposure unit (often a river basin) and the study area.
- Select a time horizon.
- Identify a preliminary range of adaptations.
- Determine general data availability.
- Determine the need for integration across sectors.

Models are often used to assess the biophysical components of a water resources assessment (hydrologic resources, water quality and aquatic ecosystem integrity) and the socio-economic components (demand from water use sectors and the water management system). Feenstra et al. (1998) provide an overview of such models.

Once an assessment method has been selected and tested and the necessary data have been collected, the key inputs and assumptions need to be formulated. Before applying a method it is necessary to develop climatic and socio-economic baseline scenarios, climate change scenarios, and assumptions about the potential for autonomous adaptation.

Possible adaptation options include:

- Modification of existing physical infrastructure.
- Construction of new infrastructure.
- Alternative management of the existing water supply systems.

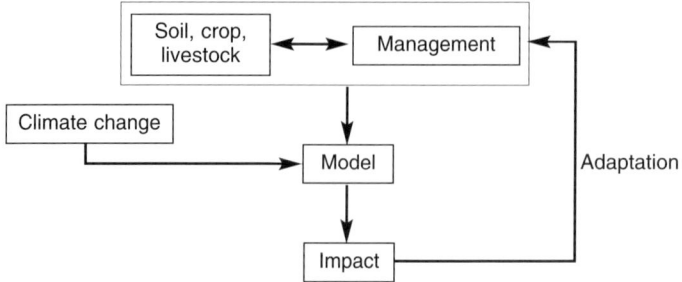

Fig. 1.4. Impact and adaptation.

- Conservation and improved efficiency.
- Technological change.
- Market- or price-driven transfers to other activities.

In Chapter 2, three water resources studies are described: in Ghana (Section 2.6), Mali (Section 2.8) and Mongolia (Section 2.9). Section 3.4 provides a synthesis of the water resources studies.

Agriculture

Agriculture, the art of cultivating the soil, growing and harvesting crops and raising livestock, combines agronomic and economic aspects of the utilization of ecosystems by a household or farmer. Agriculture is, in many countries, the most important economic activity both in terms of cash flow and of labour.

The management of soil, crops and livestock can be modelled. The impact of changing climatic conditions can be calculated using the model and based on the results changes in soil, crop and livestock management can be implemented. The adaptive capacity of a system will determine whether such adaptations are implemented. The emissions inventories (N_2O, CH_4 and CO_2) were not linked to the agricultural studies but were included in the mitigation studies.

Agriculture was part of the following country studies: Bolivia, Bhutan, Costa Rica, Mali, Mongolia, Senegal and Yemen. The task for the technical assistant was to provide tailor-made assistance building on locally available knowledge, tools and data. The assistance focused on impact and vulnerability assessment and the definition of adaptation measures.

IMPACT ASSESSMENT. A framework prepared by the IVM/UNEP handbook (Feenstra et al., 1998) was distributed among the participating groups. The NCCSAP stresses the importance of simulation models in determining the impact of climate change on primary agricultural production. The effects of changes in temperature, precipitation and CO_2 on production levels can be calculated using such models. Based on the simulation results, impacts on farm and village output and income are determined. These findings may be extrapolated to the regional and national level to assess the impact of climate change on food security. Figure 1.4 depicts the structure used in the impact studies.

From an agronomic perspective, food security starts with crop production at the field level and moves up to the farm, regional and national levels. With the changing spatial scale, the focus of research changes from an agronomic or biophysical analysis to a socio-economic or political analysis (see Fig. 1.5).

Most studies in the NCCSAP addressed issues of impact on production levels at the left-hand side of Fig. 1.5. Up-scaling these findings to a regional and national scale was in most cases difficult as data and clearly defined socio-economic scenarios were lacking.

After having determined the impact of climate change on agricultural production, measures to adjust to the changed environmental conditions needed to be defined. The effects of some technical adaptation measures such as changing planting dates, moving to other varieties or crops, irrigation and weed

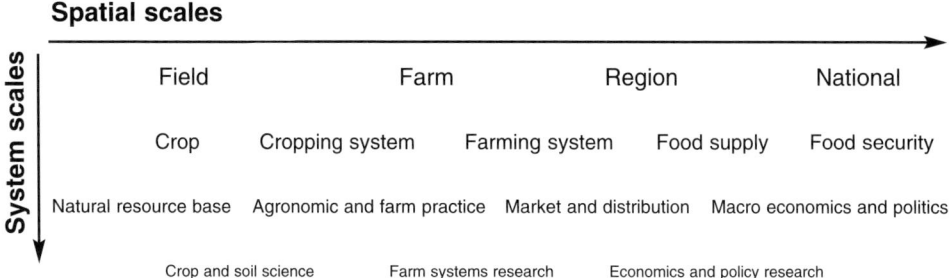

Fig. 1.5. Spatial and system scales linking crop production to food security.

control were quantified with relative ease. Again, higher level impacts, at the regional and national scales, such as changes in political and institutional settings, were addressed in more general terms.

ADAPTATION. In general, two pathways are distinguished. The first is mainly an agronomic study focusing on crop production at the field and regional level. For this analysis the results of crop simulation models are used to reveal the difference between calculated crop production under current and changed climatic conditions. As most models only combine the effects of temperature, precipitation and nutrient input for a given crop–soil combination, differences in calculated yields can be explained based on these factors. Consequently, adaptation measures based on crop simulations are also limited to these factors.

Other factors such as pests, diseases and weeds are dealt with in a semi-quantitative or more descriptive manner.

The second pathway combines the agronomic analysis with socio-economic, technical, institutional and political aspects influencing production levels or food security. This approach offers a link to current agricultural programmes. The impact of climate change on the socio-economic position of different types of farmers and options for adaptation are also evaluated.

Depending on availability of data, technology and expertise, the regional distribution of impacts and adaptation measures are part of the analysis. The context of global change expands the studies: they aim to incorporate changes in consumption patterns, demographic development and technological progress to address future demands for food, fibre, water and space. The NCCSAP also tried to look at cross-sectoral effects and competition for scarce resources, such as water.

In Chapter 2, two agriculture studies are described: in Bolivia (Section 2.2) and Yemen (Section 2.13). Subsection 3.6.1 provides a synthesis of the agricultural studies.

Forestry

Both agriculture and forestry contribute to economic development in terms of income generation and employment. In most developing countries industrial transitions will be linked to changes in agriculture and forestry. Besides changes in the production systems, these transitions will also involve changes in the industries and services based on these production systems.

Agriculture and forestry aim at producing food and fibre. To achieve this, sustainable management and use of natural resources is crucial. Quite often however, because of lack of proper knowledge or because of overexploitation, sustainable management is not achieved. The way in which land is managed determines to a large extent the quality of natural resources such as water, soil and biodiversity. Poor management is reflected in the quality of landscapes as can be seen in abandoned irrigated and eroded landscapes. It is a main priority to find a safe and responsible way to feed the growing population, which will also require the inclusion of environmental management in order to reduce long- and short-term vulnerability and to protect the natural resource base from overexploitation.

Climate change is an additional stress to systems that are, often, already under stress from other pressures. Since changes and variations in climate and other environmental factors have occurred naturally, 'adaptation' is not a new phenomenon. Both human and natural systems have repeatedly adapted to changing conditions.

Agriculture has traditionally been the key livelihood strategy for most people living in rural areas based on the main function of agriculture to provide food. Forests will not only be affected by climate change but will also provide options to generate capital via the CDM (Section 1.4.1).

Climate is not a peripheral question for development. The natural variability of rainfall, temperature and other conditions are among the main factors behind variability in agricultural production, which in turn is one of the factors behind food security. The challenge facing agriculture is to produce enough food and at the same time, to ensure that the natural resource base remains productive for the future. The main question is what the role of agriculture is in the development of the countries concerned.

Subsection 3.6.2 provides a synthesis of the forestry studies.

2
Country Experiences and Highlights

2.1 Introduction

In this chapter the key Netherlands Climate Change Studies Assistance Programme (NCCSAP) project members of most countries involved in the NCCSAP or the pre-NCCSAP studies summarize the results of the studies in their own words. Only the contributions of Bhutan and Costa Rica are not included. In the former case because the study was not finished and in the latter case because the project team separated. As can be seen from Table 2.1, 12 vulnerability studies and one mitigation study were chosen. Six of the vulnerability studies concerned coastal zones, three water resources, two agriculture and one the effects of extreme weather events.

In addition to summarizing the studies, the authors also reflect on their experiences and lessons learned. Here they describe the applicability of the methodologies used, experience with multidisciplinary research and policy implications.

The sections end with a list of follow-up studies, conclusions and an overview of the main publications that resulted from the project. If such a publication is only cited in the section itself, it is not duplicated in the References.

Table 2.1. Overview of studies summarized in this chapter.

Country	NCCSAP/ pre-NCCSAP	Subject	Section
Bolivia	NCCSAP	Vulnerability/agriculture	2.2
Colombia	NCCSAP	Vulnerability/coastal zone	2.3
Ecuador	NCCSAP	Vulnerability/coastal zone	2.4
Egypt	Pre-NCCSAP	Vulnerability/coastal zone	2.5
Ghana	NCCSAP	Vulnerability/water resources	2.6
Kazakhstan	NCCSAP	Vulnerability/extreme events	2.7
Mali	NCCSAP	Vulnerability/water resources	2.8
Mongolia	NCCSAP	Vulnerability/water resources	2.9
Senegal	NCCSAP	Vulnerability/coastal zone	2.10
Suriname	NCCSAP	Vulnerability/coastal zone	2.11
Vietnam	Pre-NCCSAP	Vulnerability/coastal zone	2.12
Yemen	NCCSAP	Vulnerability/agriculture	2.13
Zimbabwe	NCCSAP	Mitigation/energy sector	2.14

2.2 Bolivia

Oscar Paz[1], Javier Gonzales[2] and Magali García[3]

2.2.1 Introduction

The Government of Bolivia, aware of the importance of sustainable development, reoriented its national policies and included the concept of sustainability in all its actions since 1993. Until then the extractive policies in the country had led to a deterioration of the environment. One of the observed critical aspects was the occurrence of extreme events (floods, frosts, droughts) that made the community presume the imminence of future climate change. This could severely affect the productive capacity of the country and could affect its food security.

Bolivia ratified the United Nations Framework Convention on Climate Change (UNFCCC) in July 1994 and created in the same year the National Programme of Climate Change (PNCC in Spanish), within the Vice Ministry of Natural Resources and Environment of the Ministry of Sustainable Development. The intention of the PNCC is to fulfil Bolivia's obligations under the Convention. In 1999, the government ratified the Kyoto Protocol and is currently implementing Kyoto goals through the Clean Development Mechanism (CDM).

Following the above-mentioned policies, in 1998 the Government of Bolivia signed a cooperation agreement with the Government of the Netherlands through the NCCSAP. One of the main reasons for this cooperation was that Bolivia had to increase its capacity to implement UNFCCC tasks and goals, and to gather more knowledge on the effects that climate change could have on its production and its environment.

Furthermore, the undertaking of these activities would allow generation a framework to strengthen capacity in the involved institutions. In the future these could help to improve the understanding of the causes and implications of climate change. The agreement implied sound support for Bolivia for developing the fundamental elements for its First National Communication, such as the Inventories of Greenhouse Gasses (GHGs) with 1994 as a baseline year, studies on mitigation of GHG emission, studies of climatic scenarios and vulnerability and adaptation analysis of the agricultural, livestock, forest and water resources sectors. This work was developed in coordination with the State University of San Andres[4], the National Service of Meteorology and Hydrology (SENAMHI) and two NGOs[5]. Its results constituted important inputs for the development of the *First National Communication of Bolivia* (Ministry of Sustainable Development and Planning, 2000) for the UNFCCC, which was presented at the Sixth Conference of Parties (COP 6) in The Hague.

The results obtained in the different studies showed that the agricultural sector would be affected by climate change, partly because it is the economic sector that makes the least use of technology and also because it is the sector with the greatest direct dependence on climate for its productivity. In addition, climate variability affects the food and productive security of a great part of the country's population. Therefore, in this section the results of vulnerability and adaptation studies for the agricultural sector are presented.

2.2.2 Approach

The study was based on the use of the crop simulation model DSSAT3, which included the

---

[1] Coordinator of the National Programme of Climate Change (PNCC), Vice Ministry of Natural Resources and Environment, PO Box 6389, La Paz, Bolivia. Tel./Fax: 591 2 2442336, Email: oscarpaz2@hotmail.com
[2] Expert on climate change adaptation policies PNCC.
[3] Expert on vulnerability and adaptation of the agricultural sector, State University of La Paz.
[4] Institute for Agricultural Research; Institute of Hydraulics and Hydrology; Institute of Ecology; Institute of Sanitary Engineering and Environment.
[5] Initiative for the Defence of the Environment (LIDEMA in Spanish); Bolivian Academy of Sciences.

possibility of modelling the effects of an increase in the atmospheric CO_2 concentration and variations of temperature and rainfall on production in representative Bolivian agricultural areas. The model was run using data from previous climate change simulation studies. A sensitivity analysis of agricultural ecosystems to climate change was performed. Before the simulation runs, a validation was performed with available field data. For the study itself, potato (Solanum tuberosum), maize (Zea mays), soybean (Glycine max) and rice (Oryza sativa) were selected because of their economic importance and/or because of their role for food security in Bolivia. The modelling included the simulation of crop production under incremental climatic scenarios assuming maximum variations of ±4°C of temperature and ±30% in rainfall. The effects of CO_2 increase were also simulated assuming a doubling of the present concentration (660 ppm). The adaptation studies were analysed through the application of a matrix for qualitative evaluation of the effects of climate change.

2.2.3 Results

Before coming to the description of the results obtained in this study it is important to mention that Bolivia includes three different and preponderant agricultural regions: (i) the Altiplano region, located 3600 m above sea level (asl), is characterized by the production of Andean crops (potato and quinoa) that are extremely resistant to the harsh climate conditions of the zone (droughts and frosts during the cropping period); (ii) the valley region situated around 2600 m asl, producing mainly potatoes, white maize and legumes; and finally (iii) the zone of the low plains (200 m asl) where agriculture is practised extensively. Production in zones (i) and (ii) is mainly for local food consumption while production in zone (iii) is mainly for export.

Regarding the climatic characteristics and according to the studied scenarios, in the Altiplano areas as well as in the lowlands (the area with the largest exporting agricultural activity), few variations in the total precipitation are anticipated. The number of days with precipitation shows a tendency to decrease (Garcia, 2003). There is an increased occurrence of storms and also the mean minimum temperature shows a clear increase. In the Andean valleys the tendency is a shortening and intensification of the rainy period, apparently due to the combination of global climate change and local desertification. An increase of temperatures is also expected, which is likely to affect the vegetative cycle of the crops.

The modelling of the four crops mentioned in Subsection 2.2.2 was performed as follows: potato was analysed in the Altiplano and valleys, white maize in the valleys, and finally soybean and rice in the lowlands of the country (Fig. 2.1). In general, the results show that a temperature increase of 2°C would not lead to serious damage to crops, unless this rise is accompanied by an increase in rainfall. Moreover, in the case of the Altiplano, where the low temperatures slow down crop growth and extend vegetative cycles, the increase in temperature would accelerate crop growth and would finish the productive period before the start of the frost period. If a modest increase in temperature takes place without increase in precipitation, adaptive measures are needed, such as the incorporation of irrigation systems and improvement of the crop activities to maintain production. If a *reduction in precipitation* takes place, all the studied ecosystems would be affected negatively.

In the case of potato, the studies indicated (Fig. 2.1) that in the Altiplano the increase of temperature would be positive for the crop because it would accelerate its physiological activity if early frosts become less common in the zone. If the temperature increase is accompanied with an increase in rainfall, yields would be affected even more positively, by 20% on average. Due to the high altitude of the Altiplano, the CO_2 concentration is currently low (Vacher, 1998) limiting photosynthesis. Therefore, an increase in the CO_2 concentration together with an increase in temperature and precipitation would favour productivity. In the valleys, similar effects to those in the Altiplano could be perceived for potato but with more modest increases of up to 10%. However, in all cases, rainfall reduction would have serious negative impacts on

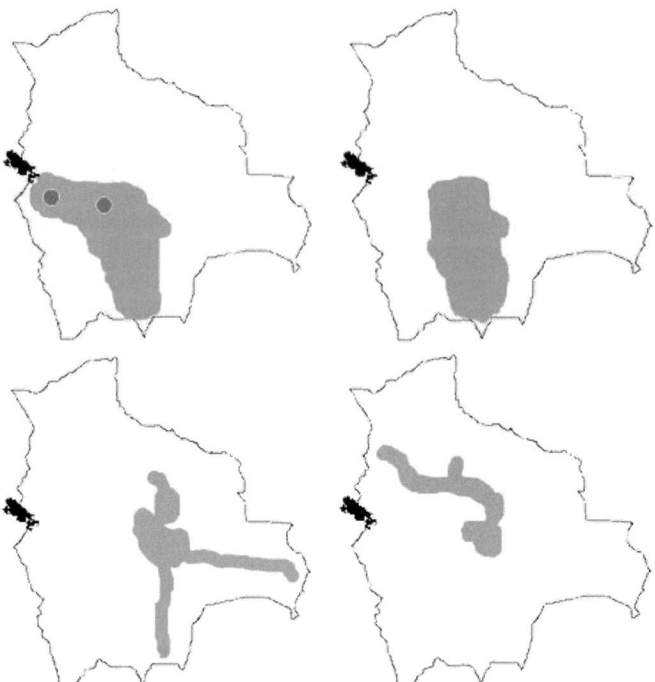

Fig. 2.1. Production areas (clockwise) of potato, maize, rice and soybean. The two round spots in the potato map indicate the case study areas.

crop production, as can be seen from Fig. 2.2.

In the valleys, irrigated white maize has a high physiological sensitivity to climate change. Yields showed a clear reduction under all scenarios predicting a temperature increase (under any precipitation regime) with an average yield reduction of 25%. A yield reduction was registered in spite of the increase in CO_2 concentration that theoretically should decrease the need for water. Apparently the sensitivity occurs due to the crop tendency to abort flowers under high maximum temperatures, which cannot be compensated by irrigation or by the increase of CO_2. The analysis of rainfed maize showed different results, as yields increased by up to 50% for all climatic scenarios (temperature increase and/or modest variations of precipitation) and doubling of the CO_2 concentration. The reason is that rainfed maize at present produces very low yields due to the intense water stress the crop undergoes. This water stress is more important than the negative effects of the high maximum temperatures. Under these conditions the doubling of CO_2 would help to reduce water stress. This gain could be maximized with the application of adaptation options such as the installation of irrigation systems and improved crop management.

Soybean modelled in the lowlands of the country showed high physiological sensitivity to climate change. For the winter (dry) season with rainfall reduction and temperature increases, the vulnerability analysis showed yield reductions of up to 45%. This is mainly caused by the severe water deficit and the shortening of the phenological phases of flowering and grain filling. In the case of an increase in precipitation, yields could be increased up to 43%, if the temperature increase is modest. When temperature increases are high, reductions in the yields will take place again. These effects cannot be mitigated by the increase of CO_2 concentration under these scenarios. In the summer (rainy) season yields are negatively influenced by the

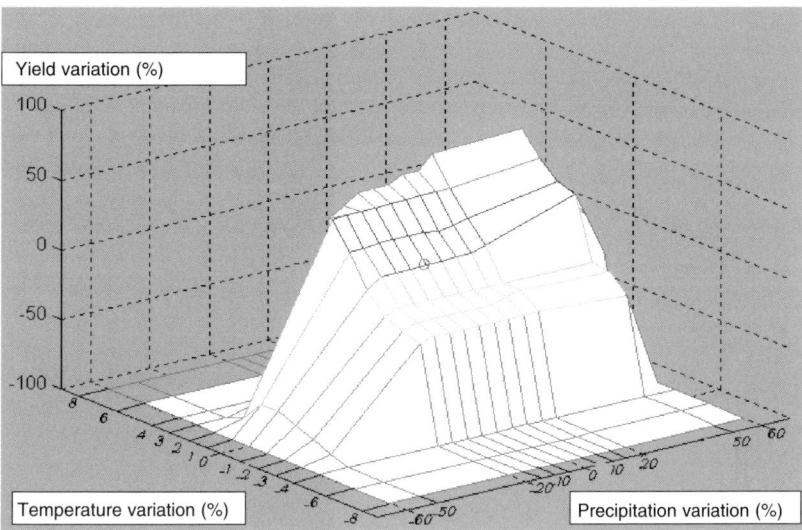

Fig. 2.2. Yield variation (%) of potato as determined by temperature and rainfall variation in the Altiplano.

temperature increase with or without increase in rainfall variation (Fig. 2.3). The maximum temperatures are much higher than in winter, which affects the flowering and grain filling phase. Changes in precipitation will not affect the production severely, as long as it does not exceed ±20% variation from the current range.

The increase of CO_2 would have positive effects on the soybean production under all scenarios, with yield increases of up to 50% attributable to the best physiological use of CO_2.

Under similar climatic conditions as the soybean (lowlands), vulnerability studies showed that rice will be little influenced by climate change, because the present conditions are optimal for the crop and small variations both in precipitation and in temperature will not affect yields. The incrementing climate

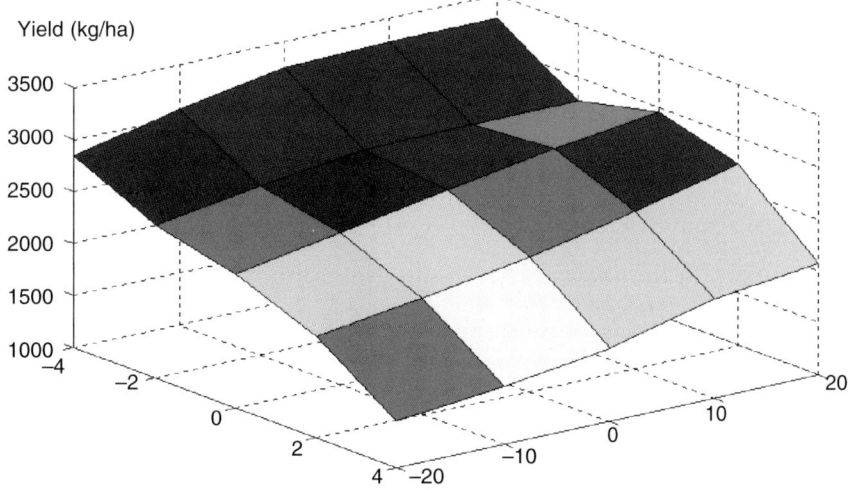

Fig. 2.3. Soybean yields under various variations of temperature and precipitation during the summer (rainy) season (at double CO_2 concentration).

scenarios showed that an extreme reduction in precipitation (up to 20%) added to an extreme increase of temperature of 3°C, would reduce the yield by 15%. The drop is not that large because the current rainfall rates are very high. The influence of the doubling of CO_2 has little effect on yield (an increase of 5%) if the precipitation levels are similar to the present ones, even under strong elevations of temperature.

The results presented above are similar to the observations made in other areas of the world: crop yields in tropical and subtropical zones will be reduced due to a combination of temperature and water stress. The results of these studies are compiled in the second and third assessment report of the Intergovernmental Panel on Climate Change (IPCC) (IPCC, 2001). Because the climate change scenarios performed in earlier studies for Bolivia indicate an average expectation of a temperature rise of 4°C and an average variation of ±15% in the mean precipitation, one may expect a reduction of yields when no adaptation measures are implemented in the highlands and in the valleys, but little variation in the lowlands.

2.2.4 Experiences and lessons learned

Although the models were calibrated before running the simulations, their application faced some problems. The most important one was the lack of input information. In developing countries such as Bolivia where research is not a national priority, extremely detailed basic information required for the models (for example a model such as CERES) is very difficult to find. Even in the case where such information exists, it is of low quality and difficult to access. For the present study, many of the default parameters for the models had to be used, since reliable information from the field was missing. Examples include photosynthetic efficiency and growth equations. In this regard, the team worked very closely with researchers from the Faculty of Agronomy of the State University of La Paz, who, where data were missing, provided criteria that would validate reasonable assumptions. In spite of the limitations, with little data and a reliable validation, the models produced outputs well correlated with the agricultural situation prevailing in the country. Since the baseline scenarios were well described it was reasonable to assume that the simulated scenarios would reflect the future situation in the country.

Due to the lack of adequate information to consider a more extensive study, the results have limited value. Nevertheless, they explore the different critical aspects of the vulnerability of agriculture to define as clearly as possible the effects of climate change. When evaluating the model results it is important to consider the different sources that may cause variation. One of these sources could be the existing large biodiversity in the country, which could influence the foreseen results especially in regions where monocropping is not a general practice. This is the case in areas where white maize and potato are cultivated.

The sensitivity analysis of the studied crops (potato, maize, rice and soybean) suggested possible phenological changes caused by the increase in temperature and atmospheric CO_2. This is not certain because the impacts might be compensated with physiological adaptation of the crops. A clear overview of the impacts of climate change on agriculture cannot be obtained yet, only the tendencies can currently be described.

For Bolivia the level of knowledge of the climate variability in the country is not yet consolidated into a solid baseline for use in these scenarios. There are only a few climate studies that describe the climatic patterns of the country. The combined effects of the latitude (tropical latitude) and physiography of the country (diverse altitude) make it difficult to make good projections for the country. In many cases the results can only be considered hypothetically.

One of the most evident learned lessons is that there is a need for better understanding of the climate variability in the different regions of the country. This is a need to perform a correct interpretation of the results obtained in all sectors. Insight into the trend of climate change can contribute to an additional understanding of the climatic baseline and future scenarios and could amplify the use of sensitivity models.

Another aspect that has not been considered in the studies, but which is absolutely relevant for understanding the impact of climate change in mountain ecosystems, is the withdrawal of glaciers in tropical mountains. This could have major consequences on local and regional hydrology and indirectly on agriculture. It should be considered as a political priority to research the impacts of the expected changes.

The simulation models used in the studies require a large amount of climatic and physiological information and this information was not always available. This meant that in many cases the work was done with default values, which seriously affects the validity of the obtained results. The use of simpler models could make the vulnerability analyses easier, without sacrificing the precision of the results. In this way a suitable balance can be found between the efficiency of the proposed models and their data requirement. On the other hand these models consider only the behaviour of the species and sectors independently of the inter-specific and inter-sector interactions, which are common in most of the Bolivian production systems. For that reason a multidisciplinary approach must be considered in future evaluations, because the interactions within productive systems might greatly influence the impacts of climate change.

The studies also favoured the increase of knowledge of the subject of climate change and of the application of simulation models, in all sectors. A positive effect is that students and lecturers of the State University learned to work with these models, which has a strong multiplying effect for the new generations of professionals. Finally, it is important to mention that the studies formed the basis of the design of policies that used to be absent in Bolivia. The studies helped in the processes of understanding the impacts of climate and global change and supported the formulation of national strategies to deal with these impacts. The PNCC takes the lead in this as can be seen from Box 2.1.

2.2.5 Follow-up research

The influence of the results of the study on the collective awareness of climate change was reflected mainly in municipalities and universities. The study results served as guidelines for field studies carried out afterwards. In many cases, the field studies confirmed the obtained results.

In order to have a better overview of the local and/or sector vulnerability, the studies must review different measurable elements of the climate such as changes in the regional hydrology and crop sensitivity. At the same time they should also evaluate the local economic, technological and institutional capacities of the population. It is evident that from the perspective of local response capacities, the results based on the use of simulation models must be complemented with studies and methodologies on the different institu-

---

Box 2.1. The impulse of the National Programme of Climate Change (PNCC).

The PNCC has promoted a gradual mentality shift in the Bolivian society about the importance of considering a detailed analysis of climate change.

The PNCC has developed a number of activities for and together with the State University. The PNCC intensified its efforts to work with the university because it considers the education of trainers as one of the best ways to increase awareness of climate change issues. In this regard, the PNCC, in the framework of the project, organized several training courses and exchange seminars with the university in its different institutes. In this way the university personnel received intensive training in the subject of climate change, in such a way that the amount of academic research related to global change has increased substantially.

The establishment of the Inter-institutional Council of Climate Change (ICCC) was initiated by the PNCC and at present the ICCC plays an important role in advising on climate change policies in the country.

Finally, the development of the National Strategy for the Implementation of UNFCCC (ENI in Spanish) is a result of the NCCSAP project.

tional adjustments needed to face the impacts of climate change (what the IPCC calls spontaneous adaptation). In this sense there have been proposals to continue with the studies made in this first stage, with the inclusion of the evaluation of local capacities and the extension of the geographic scenarios to validate and to complement the obtained data. Unfortunately the lack of national and international support has limited these efforts.

The studies also suggested exploring the technical aspects of adaptation mechanisms in more detail. From the moment the NCCSAP studies were finished, a series of subsequent studies was performed evaluating the impact of agronomic adaptation measures, to explore better-adapted species to situations of extreme events and to determine the incidence of reductive factors to productivity. Unfortunately, these studies were superficial, due to the lack of financial support. It is important that governments and international institutions prioritize those adaptation studies that will also effectively investigate future extreme meteorological events.

One of the remaining activities is to improve the understanding of climate variability in the different regions of Bolivia. A greater insight could improve the use of the incremental scenarios in these regions. Since the ending of the NCCSAP programme, advances have been made in reviewing the trends in local precipitation regimes and in some natural distortions as those provoked by El Niño. One of the deficiencies in Bolivia is the lack of real time information to be able to confront emergency situations. Given that the latter is a basic need to get precise results, the real time information network needs to be supported by local governments as well as by international institutions. To date, this has not been constant and durable.

2.2.6 Policy implications

Although the analyses of vulnerability and adaptation, carried out with the support of the NCCSAP programme, left many open questions and included various assumptions, these are the only studies on the effects of climate change in Bolivia. Therefore, they have already been included in the context of public policies. For example, the sensitivity studies suggested developing adaptation measures related to water management and exploiting the capacity of crops for extreme situations (date of sowing, selection of varieties, etc.). Some of these actions are already being implemented by national programmes, as in the case of irrigation or the production of better-adapted crop varieties. Nevertheless, in many institutions the awareness of the importance of climate change does not exist yet, and their activities do not include considerations to reduce the vulnerability to climate change.

It is important to mention that the obtained results have served as a basis for the elaboration of the National Strategy for the Implementation of the UNFCCC (ENI in Spanish) that recognizes adaptation to climate change as the central axis of the national policy. The ENI has been elaborated with the component members of the Inter-institutional Council of Climate Change (ICCC) that was structured as a direct consequence of the analysis of the results obtained in the studies. Although the ENI shows the way that should be followed for the implementation of policies, many involved actors do not yet recognize the importance of climate change. Therefore, it is necessary to increase the number of qualified people and raise awareness on the subject.

Besides the importance of the results obtained and their inclusion within the national policies, the implementation on lower governmental levels is also important. Many environmental management decisions are a municipal task, which is a result of the Law of People Participation. The central government does not include these actions within its obligations any longer. It is very difficult to include scientific results in municipal policies when the actors were not included in the definition of the research priorities and in the validation of the results. In many cases, the municipalities do not even hire technicians related to environment and/or ecology and the inclusion of climate change topics as a high priority is difficult and hardly feasible. For that reason, it is suggested that the coming research should include local stakeholders to define the actions for the implementation of adaptation and mitigation of the effects of climate change. Hence a less complex and more participatory approach is required.

2.2.7 Conclusions

In general, the results of the studies indicated an increase of temperature of up to 2°C. In the Altiplano the increase in temperature would accelerate crop growth and the productive period would be finished before the start of the frost period. If a modest increase in temperature takes place without an increase in precipitation, adaptive measures are needed, such as the incorporation of irrigation systems and improvement of the crop activities to maintain production. If a *reduction in precipitation* takes place, all the studied ecosystems would be affected negatively.

An important degree of awareness on the theme of climate change at the research institutes was created. These institutions were strongly involved in the process of analysis and knowledge of methodologies and techniques for the different components of the project. The IPCC's guidelines and methodologies for the national inventories of GHGs have been spread. The understanding of simulation models allowed analysis of the possible impacts of climate change in important sectors such as agriculture, forestry and water resources. This meant an important qualitative advance in understanding the levels of vulnerability of the different ecosystems and the strengthening of the national capacities for climate research.

The qualification of personnel for the management of the General Circulation Models (GCMs) was relatively important, because this activity showed the enormous difficulty of their application over large areas of land, such as that of Bolivia, with such a large climatic diversity. The need to count on models with a greater space resolution is obvious to improve the understanding of the possible climate changes at mesoscale level. The studies also revealed the lack of institutional capacity for climatic research by the national agency in charge of this and the urgent need for researchers and policymakers in the field of climate change to be trained, as well as the necessity of better equipment.

The studies helped to identify the country's vulnerability to climate change impacts and to strengthen the central strategic line to develop adaptation measures for all sectors, especially those related to food security, water resources and natural risks. A clearly detected limitation, which has to be worked on in the future, is the identification of adaptation actions, in a joint effort with the local communities, taking advantage of their experience.

Main Publications From the Research

Garcia, M. (2003) Agroclimatic study and drought resistance analysis of quinoa for an irrigation strategy in the Bolivian Altiplano. PhD thesis in Applied Biological Sciences, Katholieke Universiteit Leuven, Leuven, Belgium, 179 pp.

MDSP (1997) *Vulnerabilidad y Adaptación de los Ecosistemas al posible Cambio Climático y Análisis de Mitigación de Gases de Efecto Inverandero*. MDSP, VMARNDF, Programa Nacional de Cambios Climáticos, La Paz, Bolivia, 258 pp.

MDSP (2000) *Estrategia Nacional de Implementación de la Convención Marco de las Naciones Unidas sobre el Cambio Climático*, ENI. MDSP, VMARNDF, Programa Nacional de Cambios Climáticos, La Paz, Bolivia, 46 pp.

MDSP (2000) *Análisis de Opciones de Mitigación de Emisiones de Gases de Efecto Invernadero*. MDSP, VMARNDF, Programa Nacional de Cambios Climáticos, La Paz, Bolivia, 102 pp.

MDSP (2000) *Emisiones de Gases de Efecto Invernadero de Origen Antropogénico en Bolivia, año 1994*. MDSP, VMARNDF, Programa Nacional de Cambios Climáticos, La Paz, Bolivia, 131 pp.

MDSP (2000) *Escenarios Climáticos, Estudio de Impactos y Opciones de Adaptación al cambio Climático*. MDSP, VMARNDF, Programa Nacional de Cambios Climáticos, La Paz, Bolivia, 253 pp.

Ministry of Sustainable Development and Planning (2000) *First National Communication Bolivia*. Ministry of Sustainable Development and Planning,

La Paz, Bolivia, 138 pp. Available at: http://unfccc.int

Vacher, J.J. (1998) Responses of two main Andean crops, quinoa (*Chenopodium quinoa* Willd) and papa amarga (*Solanum juzepczukii* Buk.) to drought on the Bolivian Altiplano: significance of local adaptation. *Agricultural Ecosystem Environment* 68, 99–108.

2.3 Colombia

Paula Cristina Sierra-Correa[6], Francisco Arias Isaza[6], David Alonso[6] and Carlos Andrade[6]

2.3.1 Introduction

Located in the north-western tip of South America, Colombia is a country with 1642 km of coastline in the Caribbean Sea continental margin and 2188 km towards the Pacific, with approximately 52 km of coastline in the insular regions.

The evidence of and great concern related to the impacts of sea level rise (SLR) on low land coastal areas have prompted Colombia to start its own climate change vulnerability and adaptation assessment titled 'Defining vulnerability of bio-geophysical and social-economic systems due to sea level change in the Colombian coastal zone (Pacific and Caribbean) and adaptation measures'. The project used the Intergovernmental Panel on Climate Change (IPCC) methodology to evaluate general preliminary adaptation measures, based on available information, scientific and policy analysis and expert knowledge (see also Subsection 1.5.2).

The results show that the institutional, legal and organizational settings make Colombia highly vulnerable to SLR. Although more than enough legal documents related to coastal management exist, they remain isolated and sector oriented. In particular integration is lacking between execution, monitoring and enforcement levels of administration. In addition, competency and interest conflicts exist between the administrative entities as well as with the economic development sectors that benefit from coastal zones. There are also technical deficiencies within institutions to engage in the subject, and at the moment, there are no technology, information, design or execution strategies at the scientific, technical, social or economic levels.

Results from the analysis on response options and feasibility of implementation, measured by its legal, institutional, economic, financial, technical, cultural and social aspects, show that the national capacity to respond to SLR is limited. These results placed the country's vulnerability between high and critical; also, future perspectives urge the establishment of Integrated Coastal Zone Management (ICZM). The population in the Colombian coastal zone is expected to increase at a higher rate than the national average. In addition, serious plans for large-scale developments such as ports and industrial areas have been revealed to economically develop the coastal region at an increased rate.

To give perspective on the details of SLR impact, three pilot case areas were studied in the project. On the Pacific coast the Guapi-Iscuandé area and San Andrés de Tumaco were selected. Along the Caribbean the large estuarine area covers the Sinú River delta system and the Gulf of Morrosquillo. The Guapi-Iscuandé area represents a vulnerable area with a low economic capacity to finance any measures. The latter two represent areas where large-scale economic development has been planned. The case of the Gulf of Morrosquillo will be presented in more detail in order to give an example of the results.

2.3.2 Approach

The methodology used was the IPCC Common Methodology (IPCC, 1992a, 1994). This method consists of seven different and inter-related steps, resulting in an action plan providing means to mitigate the SLR impacts and identify response strategies to cope with SLR (Subsection 1.5.2). This method has the advantage of reducing costs and is useful as it

---

[6] Instituto de Investigaciones Marinas y Costeras, INVEMAR, Punta Betin, Santa Marta, Colombia. Tel.: +575 4211380 Ext. 190, Email: psierra@invemar.org.co

enables the comparison of results between countries, facilitating the augmentation of information to achieve more knowledge of global vulnerability.

The project began by defining the study area, (INVEMAR, 2003a), followed by a characterization and inventory of all aspects related to the problem (INVEMAR, 2003b). Later, scenarios were defined to quantitatively analyse the vulnerability assessments of the impacts in the natural and socio-economic systems (INVEMAR, 2003c). The next step was the evaluation of the impacts, effects and responses of the natural system on higher sea levels, coastal erosion, coastal inundation and saline intrusion and the associated socio-economic impacts to those responses (INVEMAR, 2003d). With this knowledge the different response strategies that could be used in Colombia were analysed, through a multicriteria analysis for adaptation strategies (INVEMAR, 2003e). The quantitative analysis for vulnerability assessment was then developed (INVEMAR, 2003f). Finally, the project formulated an action plan with suggestions on how to imbed proactive changes into the existing institutional framework and prioritize actions to be developed at the different government levels to ensure preparedness for SLR (INVEMAR, 2003g).

The case study areas were selected to show the practical implications of SLR and to increase public awareness about the effects of climate change. The characterization of the coasts of Colombia, regarding life quality indexes, population concentration and migratory processes, were relevant factors taken into account during the selection process. The cultural diversity in the Caribbean and Pacific region embraces different indigenous and Afro Colombian groups to whom different resources and exploitation methods are attributed, e.g. there is greater development of cattle raising, mining and industry in the Caribbean in comparison to the Pacific region. For this reason, one culturally representative area on each of the coasts was used. Below we will focus on the Morrosquillo Gulf area. It contains an important urban area and shows the possible direct impact of SLR on human settlements.

2.3.3 Results of the Gulf of Morrosquillo study

To define the study area, physical, biological, social, economic and political criteria were taken into account. There are sharp differences between the Colombian Caribbean and Pacific coasts. Because of this, the physical criterion of land altitude was chosen for the study area definition, since it is particularly dominant in the coasts' dynamic processes. Unfortunately, existent cartography lacks sufficient detail and the study area had to be delineated by the available +60 m and −200 m contour lines. A 1:300,000 scale was defined for the project.

The study area was tested for the following components: bio-geophysical, socio-economic and governance. These are needed to evaluate the vulnerability assessment due to SLR. Related effects such as erosion, salinity intrusion and inundation were also considered. Very strong assumptions would have to be made in order to extrapolate the available information to a different scale. Indicators such as demography, quality of life and economic development as well as national, regional and local governance were also evaluated in the case study areas.

The Sinú River estuary and the Gulf of Morrosquillo environmental coastal unit is shaped by a mosaic of continental, coastal, insular and marine ecosystems that spreads along the central Colombian Caribbean coast for approximately 260 km. This zone includes the coasts of Tortuguilla Island, Fuerte Island and the San Bernardo Islands Archipelago. The Gulf of Morrosquillo has an approximate area of 1000 km^2 and depths between 15 and 55 m. An extensive coastal plain limits it on the north and south with two coral terraces. The evolution and dynamics of the south-west sector, located around the Sinú River mouth, depend on the fluvial and marine processes, and are especially affected by the rate of waves and tides. The case study area covers the south part of the Gulf of Morrosquillo from the Tolú's municipal head-board to the Rada Point, see Fig. 2.4.

An important characteristic of this area in relation to the vulnerability assessment is the presence of the Sinú River delta. The basin of

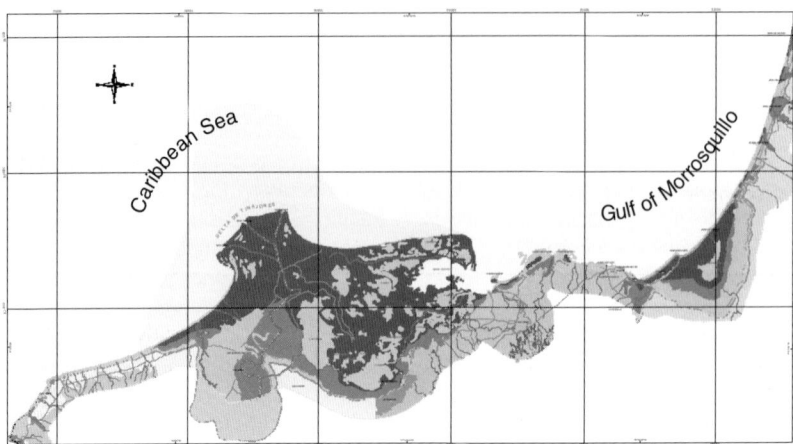

Fig. 2.4. The Gulf of Morrosquillo area.

the Sinú River is the most important of the region with an area of 13,700 km². The basin influences the water salinity in the Gulf, specifically in the south sector. During the rainy season, the salinity shows values between 26.5 and 31.8 ppt (parts per thousand) due to river discharges.

From a geomorphological point of view, the study area is divided into five regions, within which different geomorphologic units can be identified according to their origin and/or evolution: alluvial, marine, denuded, structural and lacustrine. These units are constantly changing and are influenced by marine processes, such as waves, tides and coastal currents, and terrestrial processes, including river deposits and movements. Human influence has been decisive in altering the units, primarily through structures that try to protect the beaches against coastal erosion.

The marine and coastal environments encountered in the southern area of the Gulf of Morrosquillo are mangrove ecosystems, coral reefs, meadows of sea grass and soft bottom communities typical of the continental shelf, estuaries, deltas and coastal lagoons. These environments are part of the whole geomorphologic environment of the coastal zone; they serve as a habitat for different biological resources and are used as a means of sustenance by a large part of the population of the coastal zone of the area. The mosaic of ecosystems found in this area and the dependence of local populations permitted the assessment of the vulnerability of these systems to SLR with the purpose of extrapolating the impacts and changes to similar environments of the entire Caribbean coastal zone.

The Gulf of Morrosquillo is not only important for its physical characteristics and biological diversity, but also for its socio-economic aspects and development problems. The inhabitants of the Caribbean Colombian coast have great expectations for development in the selected region, since economic zones and natural systems of recognized importance are found. These are for example Coveñas' maritime oil terminal, Tolcemento's industrial port, La Caimanera wetland, and systems of marshes and swamps of Cispatá's Bay, the Sinú River delta and Fuerte Island. Because of the strategic location of the Gulf of Morrosquillo on the Caribbean Sea, it has been the centre of a process of consolidation in economic terms and services in the territorial area through time. These characteristics have determined the importance of the region in relation to development activities (port, tourist and fishing activities among others).

The SLR impact in the Gulf of Morrosquillo was analysed for both an optimistic socio-economic scenario (low social conflict, high rate GDP growth, constant population growth) and a pessimistic one (high social conflict, low rates of consumption and inversion, low population growth rate). Table

Table 2.2. Estimated effects of SLR in the Gulf of Morrosquillo.

Year	Population affected	Agriculture (%)	Aquaculture (%)	Cattle (%)
Optimistic scenario				
2030	24,370	94	3	3
2100	44,796	32	46	22
Pessimistic scenario				
2030	23,940	94	3	3
2100	53,270	32	46	22

2.2 shows the affected population for both cases and the proportions of the impact in the socio-economic sectors.

As shown on Table 2.2, the socio-economic impacts measured on a percentage basis do not demonstrate differences between the optimist and the pessimist scenarios, due to the assumption of maintaining constant participation of the economic sectors. Nevertheless, in absolute terms, there are changes among scenarios in the temporal horizon.

In spite of these results the affected gross income is reduced by 20% for the optimistic scenario in comparison with the pessimistic scenario for the year 2030. This same analysis shows a 59% lower gross income in the pessimistic scenario compared to the optimistic scenario in 2100. As an example, Table 2.3 indicates the changes in land use at an SLR of 30 cm.

While studying the specific strategies and cost analysis for the Gulf of Morrosquillo, we found that significant measures should be taken. The main response strategies identified are:

- the formulation of an ICZM proposal in which more effective governance is achieved, especially integrated assessment with several administrative levels and sectors;
- more specific implementation of a spatial planning and an ICZM proposal to combine ecological and industrial development;
- the establishment of a new regulatory measure for design and construction of factories, houses, infrastructure, roads, etc. taking into account the potential 1 m SLR;
- road protection against erosion between Tolú and Coveñas by reducing wave impact, e.g. by the construction of a marine reef or mangroves;
- local research studies on sediment dynamics in the Gulf of Morrosquillo;
- implementation of a drainage system for rainwater in the urban area of Tolú;

Table 2.3. Changes in land use, Gulf of Morrosquillo.

Land use	Actual Use (km²)	Area lost without SLR (km²)	Annual growth rate without SLR (%)	Future (2030) Use (km²)	With 0.3 m of SLR (km²)	Land lost with 0.3 m SLR (%)
Aquaculture	5.3	0	2.2	10.3	10.3	0.2
Agriculture	61.7	0	1.5	97.2	67.0	31.1
Extractive forest	53.8	0	−0.6	44.5	44.5	0.0
Cattle	128.4	0	−1.1	92.2	91.8	0.5
Conservation of river basins	0.9	0	0.0	0.9	0.9	1.5
Fishing	1346.6	0	0.0	1346.4	NA	0.0
Urban	8.7	0	1.6	14.0	13.7	1.7
Tourism	4.0	0	−0.2	3.8	3.5	8.9
Total	1609	0	3.4	1609	1589	1.3

- a management plan for protection and restoration of river basins;
- a plan for managed (population) retreat (especially for Tolú);
- the restoration of the sediment dynamics of the Sinú River;
- the formulation of mangrove conservation plans;
- the elaboration of a strategy for aquifer protection against salt intrusion;
- the maintenance and construction of dykes.

Most importantly, the implementation of these measures will result in more effective governance of the region, permitting the appropriate use of resources in the definition of critical paths to reduce the impact of enhanced erosion due to SLR.

2.3.4 Experiences and lessons learned

One of the most recurrent problems during the development of the project was the lack of relevant, sufficient and reliable information from which to produce an analysis and apply a common methodology. Many assumptions were required to supply the information gaps or make the scarce available information useful. For this reason, the action plan developed included a section related to information gaps that outlined areas where more research was required.

We found that information of two basic types in Colombia was lacking: general information on the actual distribution of the socio-economic activities on the necessary scale of applicability, and general geographic information of the necessary scale.

The current lack of information only allowed a tentative approximation of the potential impact of SLR. Next to the lack of necessary information, existing information was not always available and some communication barriers existed between researchers, institutes and information sources (demonstrating the general need for integration and collaboration between administrative units).

The action plan outlined those information gaps together with actions/projects proposed to overcome such deficiencies. It is important to point out that some of the activities proposed had already started. Until now not enough historical data have been gathered to enable comparisons or to study the impacts.

Although the action plan was focused on information gaps, there were other aspects that involved information management and that represented significant problems as well. These included awareness of the availability of existent information, incompatibility between information systems, lack of articulation, overlap between entities that produce information, and technical limitations to producing information.

Information sources for the project included research institutes, city councils, private and public universities, territorial authorities, public libraries and consulting firms. As a result, it was a large bibliographical base that can be analysed in more detail to complement the information presented and for further contributions.

2.3.5 Follow-up research

In the future, we want to incorporate research proposals that cover the subject of SLR (impacts, vulnerability and adaptation) into the National System of Science and Technology, bringing the opportunity of more detailed knowledge of SLR impacts on particular regions. A second goal is to produce inventories on the national and regional investigation capacity related to vulnerability and adaptation capacity of socio-economic systems and natural ecosystems to SLR. In this sense, the importance of natural land-creating processes (e.g. mangroves or inter-tidal areas) will be acknowledged and studied. We will encourage the organizations responsible for national cartography to give appropriate attention to studies related to detailed cartography on coastal areas, starting with critical areas and ecosystems. We will also aim for the creation of a scientific base to generate knowledge on the marine and coastal ecosystem structure and functioning. It is very important to strengthen studies on coastal geomorphology, and strengthen the articulation between the National Environmental System (SINA) and the National System of Science and Technology Marine Sciences programme.

This effort will result in the improvement of the information supply in the national information accounts. Hence, related statistics can be applied to productive sectors, to articulate inter-institutional efforts for the creation of scenarios and to develop research proposals aimed at identifying, evaluating and prioritizing adaptation options related to climate change.

We expect that consolidated monitoring on environmental and socio-economic variables, identified as critical indicators of SLR, will be designed at the Ministry of Environment by 2005, establishing and standardizing environmental indicators on the ecosystems and marine and coastal resources. It will also state environmental and socio-economic monitoring of coastal resources to follow SLR.

We found that it is critical to develop an integrated system of coastal information exchange and processes related to SLR, in particular at the national level. Such a system would decrease the risks associated with climate change. Also important is the development of the National Oceanic and Coastal Information System established by the National Environment Policy for Sustainable Development of Oceanic, Coastal Zone and Insular Spaces (PNAOCI). This information system would serve as the baseline from which to develop plans, programmes and projects related to sustainable development of the rural oceanic, coastal and marine areas. In addition, it would help to incorporate that essential information for reducing vulnerability to SLR into the Oceanic and Coastal National Information System.

2.3.6 Policy implications

In general, the policy implications would be that the sector-oriented planning will take SLR into account, especially in those sectors directly involved in developing coastal areas, such as ports, fisheries, aquaculture, tourism and urban planning. Departments and national entities would include aspects related to vulnerability assessment and socio-economic impacts in strategic management plans and sector plans.

The PNAOCI identified seven economic subsectors and proposed specific actions in each one of them, this being the basis for adaptation programmes. Also, the National Plan for Disaster Attention and Prevention (PNPAD) established the need to include different types of risks in the sector-oriented planning risks.

The Ministry of Environment in the National Programme will take SLR into account for Coastal Zone Management. Even though in the rural areas institutions and legal instruments already exist, the programme is still emerging in these areas and urgently requires integration to strengthen the institutional mechanisms for its implementation. Our goal is that SLR and climate change risks will have to be included in any planning instrument or policy adopted by the governmental entities locally, regionally or at the national level. Solutions must be participative and many levels of coastal management will be included, not only those institutions responsible for risk mitigation and attention.

We expect that at the municipal level Territorial Land-use Plans will incorporate the subject of disaster prevention as a general rule. SLR and 'climate variability' in general have not been included in any of the local spatial plans because of the lack of awareness, experience and knowledge regarding the subject in the rural areas. Furthermore, references to and data collection of particular events such as hurricanes, flooding, tsunamis or the El Niño phenomenon are limited. This exercise forms the first step to improve that situation.

One of the first actions will be to define the desired functions of the coastal zone. Proposals for such projects will be formulated by INVEMAR as soon as possible with the support, participation and compromise of regional and local entities. Such projects will articulate and take into account the guidelines established by the PNAOCI, the PNPAD and their correspondent documents from the National Council of Economic and Social Policy, commissioned by their programmes and strategies as well as new funds.

It is well known that the best prevention and mitigation tool that can be applied to any natural disaster is to build capacity and increase knowledge and understanding in communities and populations affected or at risk. SLR is no exception. However, there are no articulated programmes for SLR in the educational system, although there have been some initiatives at the local level. The vulnerability of cultural and social aspects was ranked high due to the low quality of life in most of the Colombian coastal areas. The armed conflict affects a great proportion of the population and influences almost all economic activities. However, there is a need to educate and train the population to make them aware of climate change and related SLR effects by all possible means, using formal and non-formal education and with the support of research institutes, the Ministry of Environment and the Ministry of Education. We have also outlined the need to train the media on the coverage of scientific news and news related to natural phenomena and disasters. Two main areas of action have been included in this line:

- Public awareness: access to information needs to be given to the society. People have the right to know about the threats and vulnerabilities that exist in the places where they live or where they invest capital.
- Public participation: stakeholders, actors and populations of the coastal areas who benefit from their resources will be considered in any SLR mitigation or adaptation programme.

Coastal communities' participation will be enhanced using the present social structures. There is already legislation in Colombia's Pacific coast that promotes citizens' participation. Instruction and education need to cover all of the coastal population, including big cities and small villages, no matter how distant they are. The PNAOCI proposed a programme involving communities in the education and participation process and in actions related to prevention and mitigation of sea level change. Such a programme suggests an active interaction between users, communities and ethnic groups in coastal management by means of education, participation, land-use planning and decision-making processes. There is a need to include more sea-related information in secondary education programmes aimed to create future awareness and support of the general public for coastal and marine territories. In addition, there will be more information on the topic of 'sustainable development' both for ICZM and for territorial use in Colombia.

It is also important to improve negotiation and management capacities of Colombia with international organizations that deal with climate change and sustainable development in general. The Ministry of Environment and the Foreign Relations Office participate actively in the United Nations Framework Convention on Climate Change (UNFCCC). However, there is a need to continue negotiating international funding for research projects as well as for response strategy projects.

Furthermore, the need has been identified to intensify cooperation with neighbouring countries to develop projects for the regional evaluation of SLR vulnerability, adaptation and mitigation possibilities. In that aspect, INVEMAR and the Ministry of Environment are already contacting neighbouring countries for developing ICZM processes based on the Colombian experience.

2.3.7 Conclusions

The main purpose of the project was to define the vulnerability of the Colombian Pacific and Caribbean coastal areas for SLR and climate change. In addition, the ability to define and execute adaptation measures of the bio-geophysical, socio-economic and country capacity ('governance') was determined. The ultimate goal of the study was to better prepare Colombia for such an event. This goal was achieved.

Furthermore, an elaborate information baseline was generated that reflects the actual situation of the Colombian coast. The project developed a preliminary qualitative model that permits the detection of changes in the coastal marine ecosystems caused by eventual SLR, as well as predicting potential inundation zones and other possible scenarios. In the international field, the information provided by the study was used for the preparation of the National Communication to the UNFCCC, as part of Colombia's commitments to this international agreement.

The results lead to the conclusion that Colombia's natural systems are highly vulnerable to SLR, and that there are great gaps in knowledge of the ecosystems' condition, possibilities for recuperation and adaptation capacities. Furthermore, the analysis shows the high vulnerability of the Colombian coastal zone, regarding affected population, economic costs of possible impacts of SLR and response strategies (in terms of GDP) in the affected areas.

The analysis indicated several critical areas: the Caribbean islands – San Andrés, Providencia and Santa Catalina, the major cities and economic zones on the Continental Caribbean – Cartagena, Barranquilla, Santa Marta and Turbo, and the major cities and economic zones on the Pacific coast – San Andrés de Tumaco and Buenaventura. Taking future planned developments into consideration, another potentially critical area would be the Gulf of Morrosquillo.

The critical municipalities and their corresponding coastal environmental units will be studied in detail, followed by vulnerability assessment projects that would continue as a result or consequence of this project. These follow-up projects will be developed on a greater scale, with greater data resolution and focus on smaller areas resulting in more concrete adaptation, prevention and mitigation measures.

With the current information, it was not possible to approach adaptation measures on a national level. For each critical area, optimal and applicable solutions will be developed according to area characteristics, needs and conditions. Analysis of financial costs and benefits of mitigation and adaptation measures will be quantified for each specific site.

Finally, this investigation was responsible for offering information to the National Environmental System, which supports the implementation of the National Environmental Policy. Hence, it increased the awareness of the general public regarding the hazards related to the SLR.

Main Publications From the Research

INVEMAR (2003a) *Defining Vulnerability of Bio-geophysical and Socio-economic Systems due to Sea Level Changes in the Colombian Coastal Zone and Adaptation Measures*. Informe Técnico No. 1 Delimitación del área de studio. Instituto de Investigaciones Marinas y Costeras José Benito Vives De Andréis (INVEMAR), Santa Marta, 79 pp. (Netherlands Climate Change Studies Assistance Programme (NCCSAP): Colombia. Definición de la vulnerabilidad e los sistemas bio-geofísicos y socioeconómicos debido a un cambio en el nivel del mar en la zona costera colombiana (Caribe continental, Caribe insular y Pacífico) y medidas para su adaptación.)

INVEMAR (2003b) *Defining Vulnerability of Bio-geophysical and Socio-economic Systems due to Sea Level Changes in the Colombian Coastal Zone and Adaptation Measures*. Informe Técnico No. 2 Caracterización e Inventario. Instituto de Investigaciones Marinas y Costeras José Benito Vives De Andréis (INVEMAR), Santa Marta, 474 pp.

INVEMAR (2003c) *Defining Vulnerability of Bio-geophysical and Socio-economic Systems due to Sea Level Changes in the Colombian Coastal Zone and Adaptation Measures*. Informe Técnico No. 3 Definición de scenarios. Instituto de Investi-gaciones Marinas y Costeras José Benito Vives De Andréis (INVEMAR), Santa Marta, 41 pp.

INVEMAR (2003d) *Defining Vulnerability of Bio-geophysical and Socio-economic Systems due to Sea Level Changes in the Colombian Coastal Zone and Adaptation Measures*. Informe Técnico No. 4 Evaluación de impactos, efectos y respuestas del sistema natural. Instituto de Investigaciones Marinas y Costeras José Benito Vives De Andréis (INVEMAR), Santa Marta, 104 pp.

INVEMAR (2003e) *Defining Vulnerability of Bio-geophysical and Socio-economic Systems due to Sea Level Changes in the Colombian Coastal Zone and Adaptation Measures*. Informe Técnico No. 5 Estrategias de Respuestas. Instituto de Investigaciones Marinas y Costeras José Benito Vives De Andréis (INVEMAR), Santa Marta, 86 pp.

INVEMAR (2003f) *Defining Vulnerability of Bio-geophysical and Socio-economic Systems due to Sea Level Changes in the Colombian Coastal Zone and Adaptation Measures*. Informe Técnico

No. 6 Definición de la vulnerabilidad. Instituto de Investigaciones Marinas y Costeras José Benito Vives De Andréis (INVEMAR), Santa Marta, 41 pp.

INVEMAR (2003g) *Defining Vulnerability of Bio-geophysical and Socio-economic Systems due to Sea Level Changes in the Colombian Coastal Zone and Adaptation Measures*. Informe Técnico No. 7 Plan de acción. Instituto de Investigaciones Marinas y Costeras José Benito Vives De Andréis (INVEMAR), Santa Marta, 57 pp.

2.4 Ecuador

Hernán R. Moreano[7]

2.4.1 Introduction

The Low Guayas River Basin, located in southwest Ecuador, is the most productive area of Ecuador. It is an important area for agriculture and aquaculture, which are the leading export products, together with oil. The contribution of this region to the GDP is US$20 billion. Agriculture and aquaculture provide direct and indirect jobs to 3.2 million people of a total population of 12.1 million. The inhabitants of the basin live mostly concentrated in Guayaquil (2.2 million), Ecuador's major port and economic capital. Besides agriculture, the area holds a mangrove ecosystem of 120,000 ha and an estuarine water body of 5100 km^2. These are associated with the Guayas River and Estero Salado estuaries, connected with the Gulf of Guayaquil through the Jambeli and El Morro channels.

The climate of Ecuador's Coastal Zone is determined by the position of the Inter Tropical Convergence Zone (ITCZ), the distribution of offshore water masses and climate variability events such as El Niño Southern Oscillation (ENSO). All three are responsible for flooding of the low-lying areas in the basin and for sea level rise (SLR) within the estuaries. The last ENSO event of 1997–1998 lasted 10 months and impacts were valued at US$2.6 billion for the entire coast. This is a rough estimate made by the Latin American and Caribbean Economic Commission (CEPAL, 1998).

Climate change is a risk for Ecuador and its coastal zone and especially the Low Guayas River Basin. In 1997 the Ministry of Environment and NCCSAP agreed to conduct a vulnerability assessment study for precipitation and SLR in this area. Four Ecuadorian institutions carried out the study: Instituto Oceanográfico de la Armada (INOCAR) for SLR in the estuaries, Instituto Nacional de Pesca (INP) for impacts on mangrove ecosystems, biodiversity and the shrimp industry, Instituto Nacional de Meteorología e Hidrología (INAMHI) for precipitation assessment of the low basin hydrographic systems and Instituto de Planificación Urbana y Regional (IPUR) of the Catholic University (UCSG) for social and economic impacts in the study area. The vulnerability assessment was successfully completed 18 months later (Grupo Básico, 1999) and its products and results were presented at an international workshop held in Guayaquil in June 1999. In addition, the results were used as input in the National Communication prepared by Ecuador in 2001.

Problem definition

Under an integrated vision of resources, environment, social and economic issues, a vulnerability assessment of the Low Guayas River Basin to climate change was conducted. In this study precipitation, SLR and ENSO events were considered as a matter of climate variability and were taken into account.

The goals of the study included:

- To assess the possible impacts of climate change on the present and future situation of coastal resources and their consequences on economic and social matters.
- To identify possible attenuation measures and projects that will accommodate such measures.
- To develop planning tools for impact assessment on long-term changes of resources and their use.

---

[7] Ministry of Environment, Under Secretary of Coastal Environmental Management, P. Icaza y Pichincha, Guayaquil, Ecuador. Tel.: 593 4 2560870, Fax: 593 4 2565059, Email: hmoreano@gye.satnet.net

- To contribute to the establishment of an institutional structure for management of Ecuador's Coastal Zone.
- Capacity building in Integrated Coastal Management.
- To contribute to the National Communication of Ecuador.

2.4.2 Approach

The vulnerability assessment study was carried out using the seven steps described in the *Assessment of the Vulnerability of Coastal Areas to Sea Level Rise*, prepared by the Intergovernmental Panel on Climate Change (IPCC) in 1991 (IPCC, 1991, 1992a; Carter et al., 1994; see also Subsection 1.5.2). Steps 2 to 4 were made by the project coordinator and the working groups of institutions involved in the project, while a series of workshops was organized to define the boundaries of the study area, the climate scenarios for precipitation and SLR, formulation of response strategies, results assessment and proposal of a long-term structure to manage the Low Guayas River Basin.

The Common Methodology focused on the assessment of the impact on the natural and socio-economic system as a consequence of the physical effects of SLR and on the effects of response strategies. The vulnerability assessment starts with a delineation of the study area and specifications of SLR boundary conditions, followed by an inventory of the study area characteristics and the identification of relevant development factors related to production activities, capital investment and natural values.

After completing these three initial steps, an assessment of physical changes and the natural system response follows, which includes morphological development of the shoreline, water levels and changes in salinity. Then the response options are formulated with an estimate of their cost and an assessment of their effects, considering scenarios that represent cases showing the cost of the response options in each economic scenario (with and without development). The vulnerability assessment ends with assessing the susceptibility of changes imposed by SLR and related socio-economic impacts, followed by actions to develop a long-term basin management plan based on integration, and participatory decision making.

2.4.3 Results of the vulnerability assessment

The study area was selected because of its sensitivity to SLR. Impacts of flooding are already felt and these impacts could increase in the future as a result of SLR and changes in precipitation. The perimeter is 630 km and encircles an area of nearly 15,000 km^2 that includes 200,000 ha of agricultural fields of banana, rice and sugarcane and a population of 3.4 million people, mostly concentrated at Guayaquil. This city port is located on the upper branches of the Guayas River and Estero Salado estuaries where a series of islands and archipelagos are distributed which are covered by mangrove forest and shrimp ponds. Further south, Machala and Puerto Bolivar are the main city and port (INP, 1998). See the map of the study area shown in Fig. 2.5.

Waters of the estuaries are tide driven and their volume depends on the behaviour of ocean currents in the eastern Pacific, precipitation, freshwater river inputs caused by the southward motion of the ITCZ (Inter Tropical Convergence Zone) and warm ocean water masses driven by surface currents or by the ENSO event.

The geometry of the estuarine systems and the differences in phase with sea level make a complicated pattern of circulation, especially near the mouth area where both estuaries are connected. Horizontal salinity gradients are noticeable in the Guayas River estuary because of the freshwater inflow that fluctuates during the seasons. The maximum range of the salinity intrusion is as far as Guayaquil, it does not penetrate further because of the regulated permanent flow of the Daule River controlled by the Daule–Peripa Dam further north. Vertical gradients are almost negligible and both estuaries fall into categories 1a and 1b in the Hansen and Rattray (1966) circulation and stratification diagram.

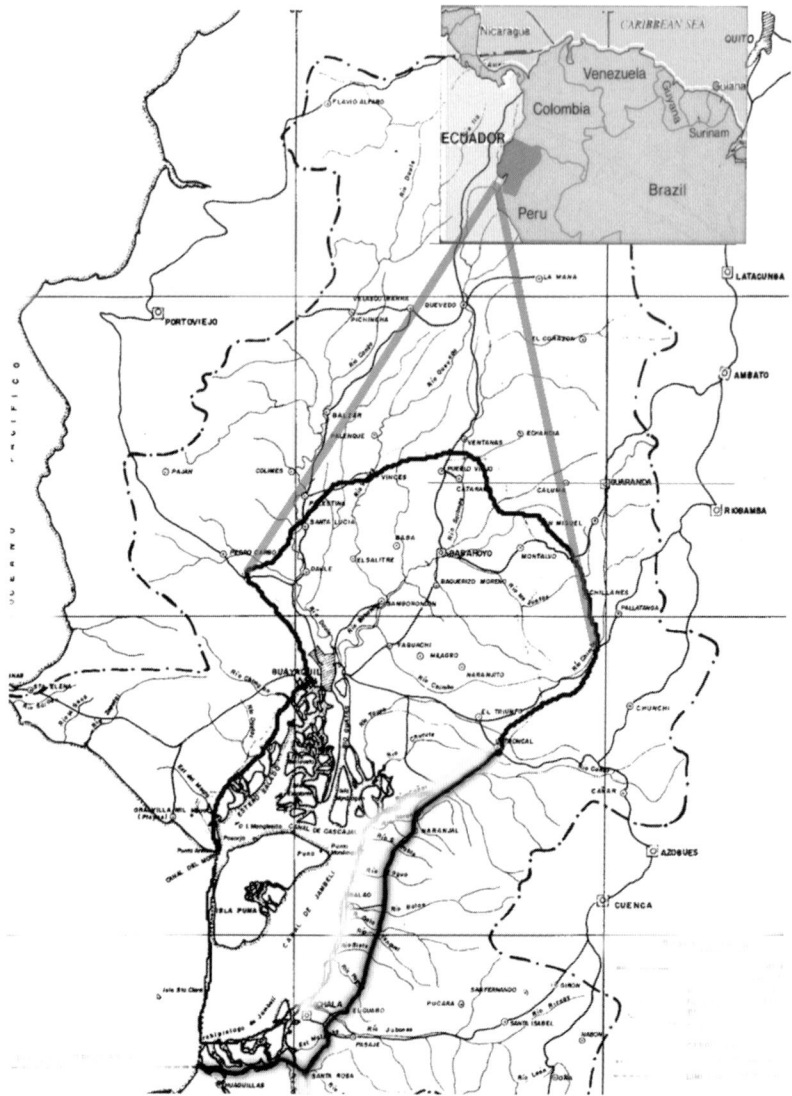

Fig. 2.5. Map of the study area in Ecuador (bounded by the thick black line).

The tidal prisms for the Guayas and Estero Salado estuaries are 4.4 billion and 1.2 billion m^3, respectively. The flushing time for the Guayas Estuary is between 8 to 13 days (Palacios, 1989) and could be less in periods of high river flows associated with El Niño. The Estero Salado upper reaches located west of Guayaquil, on the contrary, do not have freshwater inflow and flushing time is longer than in the Guayas. Going southward and closer to the mouth, the water exchange with the ocean and with the other estuary shortens the flushing time.

All three hydrographic systems are responsible for sediment transport and freshwater inflow into the study area. Calculated numbers by INAMHI are in the order of 11.2 to 22.5 million t/year and an average annual flow of 1355 m^3/s. The main contribution comes from the Guayas hydrographic system, which accounts for nearly 80% of the total.

The economic situation

The economy of Ecuador is marked by a huge external debt of US$16 billion, an annual budget with lack of resources to invest in productive sectors, low foreign investment because of lack of guarantees in the political, economic and judicial systems and high inflation and interest rates. Unemployment is about 18% and probably higher in the study area as a consequence of the last ENSO event, GDP growth was 0.8% in 1999 and there was a contraction in 1999 by about 6% because of political and economic instability.

Two sectors are responsible for one-third of income from exports: agriculture and aquaculture. Both take place in the study area. Experts expect both sectors to grow during the 21st century, but this will require new technologies and subsequent investments to obtain a higher degree of product diversification and differentiation (IPUR, 1998).

GDP for the country in 1998 was US$20 billion and is estimated to be US$9 billion for the study area. Ecuador's total population is 12.5 million people, of which 3.4 million live in the Low Guayas River Basin. The population growth rate is 2.2% per year.

Two socio economic scenarios were considered: no development from the year 1998 and development for the year 2010. These were used to make calculations on population growth and to assess the cost on capital goods impacted by climate change.

Climate scenarios

Several climate scenarios were established, and after discussions among institutional experts during the initial workshop held at Guayaquil at the end of 1997 (see Box 2.2) it was agreed to work with the scenarios included in Table 2.4.

The climate variability associated with El Niño was considered in all three scenarios, because it generates SLR of the order of 0.2 m and 0.5 m for moderate and extreme events respectively. These values are within the time series used to calculate the return periods of 10 and 100 years. The same is true for precipitation, air temperatures and sea surface temperature (SST), where values are from 50% to 80% precipitation increase, and >2°C anomalies in temperature for moderate events and 300% to 600% and >4°C anomalies for extreme ones.

As the study area is not affected by ocean storms, hurricanes and tornados, the climate scenarios do not consider data related to these events.

Results

The semidiurnal tidal wave that gets into the estuaries is deformed by the difficult geometry and hydraulic friction, at the point that tide range is 1.5 m (neap) and 2.3 m (spring) at the mouth, whereas at the head those values increased to 2.5 m (neap) and 3.6 m (spring). This means that the level of flooding is higher at the head than at the mouth (INOCAR, 1986), for this reason, the estuaries were divided into five sectors to apply levels of flooding according to location and climate scenarios (INOCAR, 1998). Although for Guayaquil the high values are applicable, average values were taken instead to establish the impact zones for each scenario. These zones are indicated in Fig. 2.6.

Capital loss, at risk and at change, was considered for each of the impact zones for the 1998 and 2010 scenarios. Capital loss

Table 2.4. Scenarios for the vulnerability assessment study for 2010.

Scenarios	Conditions
Basic (SLR 0)	No changes in sea level, air and ocean surface temperatures and precipitation
Moderate (SLR 1)	Sea level rise of 0.3 m, with an increase of 1°C in air temperature, sea surface temperature (SST) anomalies less than 1°C and a 15% reduction in precipitation
Severe (SLR 2)	Sea level rise of 1.0 m, with an increase of 2°C in air temperature, SST anomalies over 2°C and a 20% increase in precipitation

> **Box 2.2.** Choosing climate scenarios.
>
> Outputs of most General Circulation Models (GCMs) are in degrees Celsius of air temperature variation and percentage of precipitation below or above average. Unfortunately, climate in the global context and particularly in the coastal zone of Ecuador is not only related to changes in the atmosphere, but also to changes in the offshore water masses driven by surface currents. These in turn are coupled to the wind driven circulation patterns in the Pacific Ocean. One of the problems for the Ecuadorian group of experts dealing with the vulnerability assessment study was to find a way to match the scenarios chosen by INAMHI (1998) from GCMs with the behaviour of water masses and surface ocean currents in the Eastern Tropical Pacific Ocean.
>
> The proposed INAMHI scenarios were as follows (values for 2010 compared to 1998):
>
> | Scenario | Air temperature (°C) | Precipitation (%) |
> | --- | --- | --- |
> | 1 | +1 | −15 |
> | 2 | +2 | +20 |
> | 3 | +1 | +20 |
> | 4 | +2 | −15 |
>
> After discussions, the working group of experts agreed that scenario 1 matches the moderate scenario for the vulnerability assessment, assuming an increase of water transport of the Humboldt coastal current but keeping SST anomalies < 1°C, air temperatures will increase by 1°C, precipitation will be less than average by 15% and there will be an increase of sea level of 0.3 m. Meanwhile scenario 2 matches the severe one, assuming an offshore warm water mass (the source of which is the Panama Basin) is driven southward by El Niño current; SST anomalies will rise by over 2.5°C, air temperature will increase at least by 2°C, precipitation will be above average by 20% and SLR will be 1.0 m. Scenarios 3 and 4 are far away from reality because with a cool ocean it is impossible to have a 20% increase in precipitation and with a warm ocean it is impossible to expect a decrease in rainfall.

includes all flooded land and coastal habitats such as mangroves, shrimp ponds, banana, rice and sugarcane plantations, beaches and urban areas of Guayaquil, Machala and Puerto Bolivar. Losses for SLR 1 are in the order of 10% to 14% of GDP, but for SLR 2, they go up to 12.5% and 18.5% for 1998 and 2010, respectively.

Major losses are from mangroves and sugarcane plantations and associated industry. Mangroves will almost disappear in the SLR 2 scenario, this represents a loss of over US$1 billion without considering associated biodiversity, CO_2 fixation and the role of this forest in the estuarine ecosystem. Although the cost of the response strategy is high, it is distributed through time, which allows for the investment to keep this natural system alive.

In the aquaculture industry, the location of the ponds was improved after El Niño in 1982–1983. Hence the impact of flooding will be less severe. Furthermore, the increase of

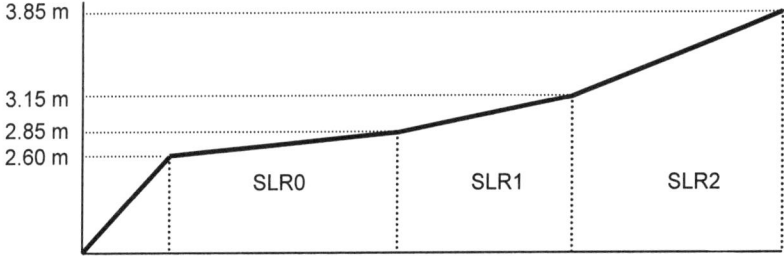

Fig. 2.6. Impact zones related to average levels of flooding for each climate scenario.

the SST is favourable for shrimp farming, increasing production and productivity. This was the case for 1997–1998, when shrimp exports were the highest ever. After 1998 the export collapsed dramatically, to 30% of the 1998 levels because of the white spot virus.

Values at risk include people and capital values facing the probability of flooding in the same areas. It is estimated that between 17,000 and 370,000 people would have to be evacuated and 75,000 to 2,000,000 people are at risk. Capital values are estimated as one to three times the GDP for the different scenarios. Values at risk include mostly financial damage to each of the sectors and values amount up to three times the GDP in the worst scenario.

Freshwater availability for all three scenarios is in the order of 35 to 45 billion m^3/year, but these numbers increase substantially during an ENSO event and especially during the extreme ones which last 10 months. Water demand by population, irrigation and industry in the same scenarios are 6 to 8 billion m^3/year, so the basin has plentiful water resources. However, there is a need to conserve these resources for the dry season. This is programmed in the basin development project through the construction of dams.

Two response strategies were considered for the vulnerability assessment study:

1. **No measures**: no actions are taken at all to face SLR and changes in precipitation due to climate change. Of course, the costs of this strategy are zero, but large costs associated with damage are very likely to be incurred.
2. **Total protection**: adaptation measures are considered and implemented to reduce losses and risk in each impact zone in order to protect people and capital distributed in each zone. The first adaptation measure that could be taken in the Low Guayas River Basin to decrease the effects of changes in climate and variability is the implementation of the development plan for the basin made by the Comisión De Estudios Para El Desarrollo De La Cuenca Del Rio Guayas (CEDEGE) (CEDEGE, 1992). Coastal defences are considered for shrimp ponds, beach nourishment for the Jambelí beach and hydraulic refills for urban areas. The vulnerability assessment study considered that the development and integrated management plan of the study area are responsibilities of the Steering Committee of Provincial Development, created by the Decentralization in Social Participation Act of 1997. The Committee is the central authority for planning and evaluation, while the implementation of projects and activities will be the responsibility of local institutions such as counties or municipalities. These have the capacity to put plans that are agreed upon with local stakeholders into action. The scientific community is represented in specific working groups of the advisory board to provide the necessary social, economic and environmental information for planners and project makers.

The combination of development and integrated management in the Low Guayas River Basin gives the study area a particular strength because each component works independently and within its own framework. They are all interdependent because they support each other with delivery of data for planners and project makers while these in turn contribute by supplying resources and asking new questions of the researchers.

2.4.4 Experiences and lessons learned

The Basic Group coordinated the work of experts from Ecuadorian research institutions. This showed that knowledge and data were available to carry out the project. It was possible to design adaptation strategies to reduce climate change and climate variability impacts on the basis of a multidisciplinary discussion that allowed integration of use and experience of participants. Unfortunately, the Basic Group finished its work in 1999, just a few months after the study was completed. Since then, matters related to climate change have been the responsibility of the Coastal Climate Change Coordinator.

The methodologies and models used provided valuable results on SLR and river flow predictions. The involved experts were pleased to transfer the knowledge of these techniques and fortunately there was enough data to feed the models. The model outputs appear to be reliable. After finalizing the project, the working group did not have the opportunity to proceed with a new study with a more multi-disciplinary approach. The people working on the project learned a lot during the 18-month period, not only about technical matters, but also about social skills and teamwork. These important achievements are kept alive by the work of the Coastal Climate Change Coordinator.

The group of experts agreed that the seven steps to the assessment of the vulnerability of coastal areas to SLR were coherent and understandable. The difficulties arose when discussing scenarios and matching outputs of the GCM with the offshore local conditions (see Box 2.2). Data to delineate characteristics of the natural system were available except data that related to erosion and accretion processes, subsidence values and groundwater sources within the estuary. It was not the same with the socio-economic system: the cost of the unit area of natural ecosystems, such as mangroves, and productive sectors, such as rice, banana, sugarcane and shrimp ponds, were especially difficult to find. This is due to the difficulty of valuing ecosystems, and the great differences in infrastructure and technology that each farm uses. After analysis and discussions, it was agreed to use average values, which allowed the continuation of the vulnerability assessment.

The response strategies were easy to identify, because most of these strategies were considered in the Low Basin Development Project prepared by CEDEGE. The main problem was to estimate the cost of the remnant infrastructure to be built in the future because of the economical instability of the country.

In the last decade, governments in Ecuador initiated a decentralization process and there is an interesting legal framework to manage coastal zones or river basins, so, it was not a problem for the group of experts to find a way within such a framework to deal with the management of the LRGB. The main problem is that the political motivation for making decisions is lacking and therefore the proposal has not been approved.

2.4.5 Follow-up research

Some months after the vulnerability assessment was completed, impact assessment and adaptation strategies were reviewed by the same group of experts. Once the final report was ready, the vulnerability assessment findings and policy recommendations were used in the National Communication. Since then, a series of seminars and workshops has been held along the coastal zone to increase the awareness of local authorities, stakeholders, municipalities and civil society about the climate change risks and results of the vulnerability assessment study. Funding for this work was provided by the responsible Ministry and by international donor agencies.

The Low Guayas River Basin is not the only estuarine area in Ecuador, there are three medium and 70 small estuaries associated with river basins and coastal valleys, city ports and beaches. On the Galapagos Archipelago the results of the climate change studies need to be communicated to the local stakeholders. This will help to empower them to implement adaptation strategies and to strengthen policies already included in the National Communication. These policies were drawn up by the government to reduce the vulnerability to climate change, as described under Article 3 of the United Nations Framework Convention on Climate Change (UNFCCC).

Many universities and faculties have already included the issue of climate change in the undergraduate and postgraduate curriculum, through seminars and workshops participation of both the national and the coastal coordinators is always welcomed.

2.4.6 Policy implications

The vulnerability assessment led to the following policies to reduce the vulnerability of the Low Guayas River Basin to climate change.

Within some years, the infrastructure of dams, irrigation and bypass channels, water transfer projects, hydroelectric plants and drainage systems designed by CEDEGE and Programa Regional para el Desarrollo del Sur del Ecuador (PREDESUR) described in the development plans for the study area will be completed.

The following projects of CEDEGE have already been completed:

- bypasses of the Bulu Bulu and Chimbo rivers;
- irrigation and drainage system of the Catarama River;
- the transfer of water from the Daule River to the Peninsula de Santa Elena;
- the Daule Peripa to La Esperanza and Poza Onda project;
- the 210 MW hydroelectric plant.

Various PREDESUR drainage and irrigation projects have also been completed. These were related to the Marcabelí Dam in El Oro province. The infrastructure will reduce the risk of the study area to climate change and climate variability events such as El Niño.

The conservation of mangrove forests as an instrument to conserve the estuarine environment and to preserve biodiversity associated with these ecosystems is implemented in the area. Mangrove cuttings are reduced as a result of a series of new regulations issued by the government through the Ministry of Environment. Cutting these forests is now a crime punished with fines and imprisonment. There is better control by authorities led by the Navy (Coast Guard and Port Captains) who work together with the 'Law Enforcement Unit'. The local communities are also empowered to defend the mangrove forest and its resources.

During the last El Niño event, the shrimp pond owners in the estuaries learned about new techniques to make the levees stronger to reduce risk of SLR provoked by El Niño and of course this will be useful for SLR as a result of climate change.

Adaptation strategies should consider integrated basin management for the development of the study area. The Ecuadorian government has applied integrated basin management in six coastal areas in the past decade. This is done through the Coastal Resources Management Programme, where civil society, combined with the public and private sector, have organized themselves. Through the programme they coordinate decisions on distribution rights, obligations and rights for the use of the shared coastal resources. This is implemented through the preparation and formal adoption of a management plan for each Special Management Zone, but it is not functioning yet.

The Decentralization and Social Participation Act is supporting this policy, because it leaves the responsibility to make decisions on development to the local civil society and stakeholders. On this basis, many municipalities have prepared management plans, drawn up through local participation. These management plans have clear goals and strategies on issues related to climate change. Furthermore, the Decentralization and Modernization Act transfers competences from the central government to the local level. For example, environment has been included in county agendas. Furthermore, according to the Environmental Act, counties must create within their structure an Environmental Management Unit (EMU).

The methodology of the seven steps allowed the definition of the problems of the Low Guayas River Basin and climate change and supported them with background data and analysis. Then some policy options were developed and criteria and weights were established to evaluate these options. This helped to make decisions based on best policies which will be adopted and applied in the study area. The best policies defined address both climate change and climate variability.

2.4.7 Conclusions

The vulnerability assessment showed the vulnerability of the Low Guayas River Basin to climate change and climate variability. The worst-case scenario (SLR 2) indicates that capital at risk and capital lost together amount to three times Ecuador's GDP of 1998. Adaptation measures to reduce the costs of the impacts of climate change amount to 10% to 20% of the 1998 GDP. That is if the measures are taken now, the cost will increase substantially if they are left for 2010. The cost–benefit relation is favourable to implement the measures now. The main constraint for implementing response strategies for SLR and precipitation changes is that the economy of Ecuador cannot bear the investments.

The Coastal Resources Management Programme, a government structure in charge of applying Integrated Coastal Management (ICM) along the continental coast of the country, is aware of climate change matters and the vulnerability assessment committees and municipalities that work with the programme within the Special Management Zones, where ICM is applied.

The Basic Group, the Coastal Climate Change Coordinator and experts of participant institutions have the knowledge and experience to focus the investigations towards the issues of impacts and adaptation.

The proposed management scheme for the Low Guayas River Basin is based on local participation in decision making. However, it is not functioning yet. This is due to the period of political and economic instability after the research of the Basic Group. Later, the Environmental Act was approved and obligates municipalities to make an integrated plan in cooperation with the undersecretary of Coastal Environmental Management, this is also the case for the Low Guayas River Basin.

Main Publications From the Research

CEDEGE (1992) *General Plan for Development of the Low Guayas River Basin*. Comisión de Estudios de la Cuenca del Rio Guayas (CEDEGE), Guayaquil, Ecuador, 75 pp.

CEPAL (1998) *Ecuador: Evaluación de los efectos socioeconómicos del Fenómeno del Niño en 1997–1998*. Informe Comisión Económica para la América Latina y el Caribe (CEPAL), Santiago, Chile, 34 pp.

Grupo Básico (1999) *Evaluación de la Vulnerabilidad de la Cuenca Baja del Río Guayas al Levantamiento Acelerado del Nivel de mar*. Report prepared by the coordinator of the Vulnerability Assessment of the Rio Guayas Low Basin to Accelerated Sea Level Rise, Guayaquil, Ecuador, 200 pp.

INAMHI (1998) *Vulnerability of Water Resources of the Low Guayas River Basin to Climate Change*. Instituto Nacional de Meteorología e Hidrología, Quito, Ecuador, 120 pp.

INOCAR (1986) *Oceanographic Studies for Dredging the Estero Salado Navigation Channel*. Instituto Oceanografico de la Armada, Guayaquil, Ecuador, 250 pp.

INOCAR (1998) *Sea Level Rise and Salinity Intrusion*. Instituto Oceanografico de la Armada, Guayaquil, Ecuador, 80 pp.

INP (1998) *Estuarine Natural System*. Technical Report, Instituto Nacional de Pesca, Guayaquil, Ecuador, 80 pp.

IPUR (1998) *Socio-economical Study*. Report, Instituto de Planificación Urbana y Rural, Guayaquil, Ecuador, 40pp.

2.5 Egypt

Ayman F. Abou Hadid[8] and Helmy Mohamed Eid[9]

2.5.1 Introduction

Egypt depends on the Nile River as the primary water source. Its large traditional agricultural base and its long coastline already undergo both intensifying development and erosion. The rapid growth of the country's population, the economic stress of reliance on food imports and the limited area for agriculture require Egyptians to find new ways to increase agricultural productivity and water resources. The Egyptian coasts extend over more than 3000 km and include coasts on the Mediterranean and the Red Sea in addition to a number of lagoons situated along the Nile Delta coast and to the east of the Suez Canal. Coastal lagoons and lakes are, in general, zones of high productivity and are extremely sensitive to disturbances (Fig. 2.7).

This, combined with the country's economic situation, makes the coastal zones and water resources in Egypt vulnerable to climate

---

[8] Central Laboratory for Agricultural Climate (CLAC), Ministry of Agriculture, 6 El Noor Street, Dokki, Cairo, Egypt. Tel./Fax: +202 7490053, Email: ruafah@rusys.eg.net
[9] Soils, Water and Environment Research Institute (SWERI) ARC, Ministry of Agriculture, Egypt.

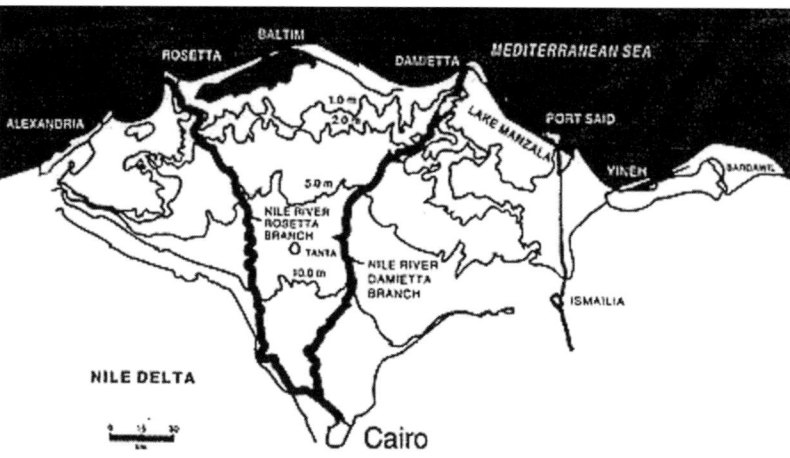

Fig. 2.7. Topographical map of the Nile Delta (Source: Nicholls, 1991).

change and particularly to sea level rise (SLR). Many of the deltas would be subject to inundation in the case of SLR. On the global scale, significant reduction of coastal wetlands appears likely and may result in negative impacts on other resources, such as fisheries. Countries with large agricultural populations in deltas, such as Egypt, are particularly vulnerable. A decrease of the Nile River budget and flow rate (as predicted by some models), could result in an increase of the rate of erosion and saltwater intrusion at the Northern Delta coast, affecting agriculture and other sectors.

These effects, and a lack of accurate up-to-date information on elevation and socio-economic characteristics of the coastal zone, were selected for a more in-depth study on the possible impacts of climate change. The study included a quantitative vulnerability and adaptation assessment covering all vulnerable areas for each of the coastal governorates of Egypt. The study also included upgrading the existing Geographic Information System (GIS) by using high-resolution satellite images to accurately identify vulnerable areas, changes in sea elevation and socio-economic characteristics following the adoption of the law 4/1994 (Legislation for the Egyptian Environment Protection including coasts), updating land subsidence rates by installing tide gauges over vulnerable shores and upgrading the digital elevation models (DMD).

The study aimed to survey processes and problems along the coastal areas of Egypt, to define the 'degree of vulnerability' of the coastal zone of Egypt and to start discussing integrated coastal zone management. Examples of coasts of varying vulnerability were discussed and possible future extension of studies to address some of the unsolved problems were identified.

2.5.2 Approach

Several general assessments of the possible impact of SLR on the Nile Delta coast were carried out in the 1980s and 1990s (e.g. Broadus et al., 1986; Stanley and Warner, 1993). An SLR vulnerability assessment finalized in 1992 (Delft Hydraulics, 1992) included a preliminary assessment of socio-economic impacts on Egypt's northern wetlands and fisheries. In one scenario, aquaculture development in the northern lakes would have a large effect on fish production potential.

The River Nile Delta of Egypt is considered one of the areas most vulnerable to SLR. Large areas south of Alexandria, Rosetta and Port Said are considered highly vulnerable. Rosetta is the anglicized name of the city of Rashid, a port city on the Mediterranean Sea in Egypt, located 65 km east of Alexandria. Excessive erosion rates have been observed at the Rosetta promontory tip, due to cessation of sediments after building the Assuan Dam in the River Nile about 1000 km to the south.

The region surrounding the city is well known for its water logging and water bogging problems. Current problems in the coastal zone of Port Said are beach erosion, pollution, subsidence and SLR. Previous studies on Alexandria (e.g. Emery et al., 1988; El-Raey et al., 1999) indicated moderate SLR ranging from 0.5–0.8 mm/year (see Fig. 2.7). A few other studies used a short time series of annual tide gauge records to report an uplift of Alexandria at a rate of 0.7 mm/year due to geotectonic uplift.

The vulnerability and adaptation assessment described here was carried out using the Intergovernmental Panel on Climate Change (IPCC) Common Methodology including the Seven Steps Method: define problem, select method, test method sensitivity, select scenarios, assess biophysical impacts and socio-economic impacts, assess autonomous adjustments and evaluate adaptation strategies (IPCC, 1994; El-Raey et al., 1997; see also Subsection 1.5.2).

The following methodology was used for assessing baseline adaptation and future adaptation options:

1. Identify and prioritize vulnerable sectors and localities based on the country vulnerability assessment;
2. Evaluate the socio-economic baseline scenario (i.e. existing and already planned policies and measures based on technical and economic efficiency, social goals and environmental sustainability) using multicriteria analysis;
3. Evaluate the physical baseline scenario (i.e. the impact matrix) using multicriteria analysis;
4. Evaluate the identified adaptation options to cope with SLR;
5. Compare the baseline scenario and the SLR adaptation options, and identify the effectiveness of the adaptation measures.

A multidisciplinary team of researchers carried out this assessment. This method also enabled assessment of adaptation capacity development with time.

2.5.3 Results of the vulnerability and adaptation assessment

The IPCC method results indicated that agriculture, industry and tourism would be most vulnerable to SLR. However, employment analysis indicated a different order of vulnerability: industry, tourism and agriculture. A more detailed assessment for the cities of Port Said and Rosetta indicated that, without adaptation, Port Said is at risk of significant impact on various sectors (El-Raey et al., 1997) while Rosetta is at risk of significant land losses and effects on historic and archaeological sites, employment and agricultural lands. The study showed that not all the coastal zones of Egypt are vulnerable to SLR to the same extent. Identified vulnerable areas along the northern coast included:

- narrow sand barriers separating the Mediterranean from the coastal lagoons, such as barriers of Burullus, Manzala and Bardawil lagoons;
- the flat and low-lying coastal plain (less than 2 m high), such as the backshore area south of Abu Quir headland, the Manzala lagoon area, and El Tineh plain, east of Port Said;
- deltaic coastal plain areas affected by vertical land movement (subsidence) due to sediment compaction, dewatering and tectonics.

Identified less vulnerable areas included coastal plains backed by high-elevated features, such as carbonate ridges and sand dunes that would act as a natural defence system against SLR. Examples include:

- the rocky carbonate ridges backing the western beaches of Alexandria and the local rocky limestone islets off the waterfront of Alexandria city, acting together to protect the low-lying areas south of the city and west of Abu Quir headland; and
- the coastal dune systems backing the coastline of Abu Quir bay, Gamasa embayment and El Tineh bay.

Areas experiencing tectonic uplift, such as the west coast of Alexandria and the Red Sea, will be probably less affected by the SLR than other areas.

Low-lying coastal zones experience a moving coastline, i.e. accreting beaches, even though they are subsiding. This may be due to the tourism activities across the coast where new beaches were built. These areas are not vulnerable to SLR if the rate of accretion exceeds or at least balances that of relative

SLR. More research is needed to identify levels of vulnerability and high-risk areas. Although the length of the Nile Delta is undergoing regional retreat, there are a number of localities experiencing accretion. On the whole, the eroded sand is carried along the shore eastwards or locally to the west and subsequently is deposited in areas of beach accretion within saddles or embayments between promontories. Such areas are the central zone of Abu Quir bay, Abu Khashaba (west of the Rosetta estuary), Gamasa embayment and El Tineh bay.

The assessment showed that, without adaptation, millions of people would be at risk of losing their homes and needing to migrate due to flooding in a 0.5 m SLR scenario by the year 2050 (Delft Hydraulics, 1991; Sestini, 1992; El-Raey et al., 1995). Without adaptation, the capital at risk was estimated at US$2.5 billion (at 1992 prices) under the same scenario. Additionally the estimated loss of productive land would have serious implications on job opportunities and food availability. It was estimated that 15% of the arable delta land would be subject to inundation, extending as far as 20 km inland from already existing coasts, with a 1 m SLR scenario by 2100. Additionally, land productivity would suffer due to saltwater intrusion effects up to the 2 m contour line, which is 30 to 60 km wide. This belt includes important cities such as Alexandria, Port Said, Kafr El-Dawar, Rosetta, Damietta, Mataria and Manzala (Sestini, 1992; El-Raey et al., 1995).

2.5.4 Adaptation

Adaptation baseline policies and measures are defined as the set of policies and measures already taken by various concerned authorities and NGOs to minimize adverse impacts of SLR from the perspective of the precautionary principle. This also includes past experience with adaptation such as technical means of enforcement, assessment and feedbacks. The RIAM software represents a comparison between baseline conditions and options for adaptation for a particular sector at a particular location (Nicholls, 1995). Aggregation of such results is also possible.

The final results could be summarized here in the following points:

1. The most severely impacted sectors are agriculture, industry and tourism. However, employment analysis indicates that the most impacted sectors are industry, tourism and agriculture, respectively;
2. The study identified a number of policy options that can be undertaken and they are: beach nourishment with hard structures and this option is already being done, application of Integrated Coastal Zone Management (ICZM), and land use change either with change to less vulnerable lands or to another land use such as aquaculture;
3. It is important to establish an appropriate monitoring system to ensure compliance with regulations and implementation of adaptation measures.

In any adaptation plan, a survey of baseline adaptation policies, measures, environmental conditions, available technical tools and past experience is necessary to ensure suitability of the adaptation measures to be taken. It was suggested that a strategic environmental impact assessment must be carried out for any policy of adaptation and an environmental impact assessment of any measure. A survey of existing policies and measures, taken by various authorities, was carried out. An assessment of measures taken, based on the above mentioned attributes, was judged by a multidisciplinary team.

A set of adaptation measures involves the management of low-lying lands on the northern fringe of the Delta, where the consequences of SLR (submergence and salinization) are likely to cause serious damage. Some of those lands cannot be used for agriculture, and the limited amount of water made available should consequently be delivered to the irrigation of new cultivated lands outside the New Valley and Delta.

2.5.5 Experiences and lessons learned

The lessons learned from the assessment are:

1. The importance of maximizing the role of the national institutions. National monitoring programmes and networks for collection and

analysis of coastal indicators necessary for vulnerability assessment are needed. A network of decentralized geographic databases based on remote sensing and ground based monitoring should be developed and updated regularly. All these activities require strong institutions.

2. Scientific capacity has been built among the research team members involved in the project because of the many activities that have been carried out. Field surveys, workshops and multidisciplinary research team meetings (scoring, discussions with different stakeholders, etc.) have led to the building of scientific capacity. Also, the tools used for the study implementation have increased capacity.

3. Sharing knowledge and information and raising awareness between different stakeholders is important for building scientific capacity and the formulation and evaluation of adaptation options.

4. The complexity of issues in vulnerability and adaptation assessment calls for well coordinated multidisciplinary research teams.

5. Methodologies used in studies need to be in line with policy needs. The issue of adaptation is very politically sensitive. Hence, proposals for solutions to already existing problems and measures to increase the adaptive capacity must go through the appropriate government channels.

2.5.6 Follow-up research

A number of follow-up activities have already been carried out. For instance, Abou Hadid et al. (2003) carried out a more detailed assessment of the impacts on agriculture. They reported that the potential impact of climate change could lead to a decrease in national production of many crops (ranging from −11% for rice to −28% for soybeans) by the year 2050 compared to their production under current conditions. The study included simulation of on-farm adaptation techniques such as the use of alternative crop varieties, optimization of the planting date, increasing water and/or nitrogen fertilizer use as well as modifying plant population density in the field. These on-farm techniques, which may imply low additional costs to the agricultural system, are compensated partially or fully by increased yields.

It should be mentioned here that El-Raey and Frihy carried out another study on climate change scenario development using MAGICC/SCENGEN in 1999 (Eid et al., 2001).

The following follow-up activities are still required:

- Carry out a complete quantitative assessment of the vulnerability of the coastal zones applying the seven steps IPCC methodology, and using the latest data, particularly that obtained by recent satellite images.
- Develop and implement national monitoring programmes and networks for collection and analysis of coastal indicators necessary for vulnerability assessment.
- Include socio-economic issues and link the agriculture and water sectors in future coastal zone vulnerability and adaptation assessments.
- Set up and regularly update national databases with all available data for and from climate change studies.
- Establish a network of tide gauges in sensitive areas, using recent systems applying Global Positioning System (GPS) and satellite technology, to provide simultaneous data on land subsidence, waves and sea levels, particularly in vulnerable sites.
- Implement and monitor integrated coastal zone management including all relevant systems and areas.
- Study possible regulatory incentives and legal options for tourism in the Egyptian back zones.
- Carry out capacity analysis and detailed vulnerability assessment for tourism in the Egyptian coast. Improved management of tourism is considered a high priority in view of the expected increase in pressures on some specific coastal sites.
- Survey and evaluate existing and planned policies and measures taken by various coastal authorities. This assessment should be based on net benefit, effectiveness, costs, first and higher order impacts, fairness to stakeholders, etc. The Coastal Protection Authority in Egypt has published many of these measures, and publications could be collected and analysed.

Although policymakers perceive adaptation to climate change as an important issue they do not have funds to support research in this area because of other pressing problems in the country. Hence, follow-up research in this area depends on donor funding.

2.5.7 Policy implications

The present study (Eid et al., 2001) has been included in the National Action Plan (EEAA, 1995) and the Initial National Communication (EEAA, 1999) to the United Nations Framework Convention on Climate Change (UNFCCC). Both are supporting attention for adaptation to climate change in the coastal, agriculture and water resources sectors.

Actual policy implications are:

1. **Beach nourishment and groins**. Beach nourishment strategies include depositing sand on the open beach and beach scraping, building artificial dunes as storm buffers and beach sand reservoirs, and laying pipes underneath the beach to suck in the water and trap sand. Groins trap the sand that covers the beach.
2. **Breakwaters**. Breakwaters are hard structures used to reduce the wave energy reaching the shoreline. The net benefit of this strategy is only along the coastline. It is the best available tool for protection of the lowland areas.
3. **Application of Integrated Coastal Zone Management (ICZM)**. For the Red Sea area as an example, the plan is structured around distinct components, including integrated coastal marine planning requirements and approaches, sustainable tourism, risk assessment and management, information management, environmental awareness and preliminary zoning proposals.
4. **Land use change**. This could mean moving productive agriculture to less vulnerable land and changing the land use of the vulnerable area to, for example, aquaculture.
5. **Legal Development Regulation**. This involves taking legal or regulatory actions to restrict development or prohibit development in a hazard-prone area. In Egypt, the regulation is effective only for private companies.

2.5.8 Conclusions

Sea level rise could have major impacts on coastal systems in Egypt (Delft Hydraulics, 1992; Sestini, 1992; El-Raey et al., 1995). The study indicates that, without serious adaptation measures, millions of people could be at risk of losing their homes. Even with the 0.5 m SLR scenario, the potential damage to the beaches and installations and the loss of productive land are predicted to have serious implications for job opportunities, food availability and population movement.

Main Publications From the Research

Abou Hadid, A.F., Medany, M., Eid, H.M., El-Marsafawy, S.M., Ouda, S.A., Ainer, N.G. and Abdel-Hafez, S.A. (2003) *Assessments of Impacts, Vulnerability and Adaptation to Climate Change in Agriculture and Water Needs in Egypt.* Progress Report of the Egyptian Case Study of the AIACC AF90 Project, Central Laboratory for Agricultural Climate (CLAC), Egypt, 13 pp.

Delft Hydraulics (1991) *Implications of Relative Sea Level Rise on the Development of the Lower Nile Delta, Egypt.* Pilot study for a quantitative approach. Final report. Delft Hydraulics, Delft, The Netherlands, 300 pp.

Delft Hydraulics (1992) *Implications of Relative Sea Level Rise on the Development of the Lower Nile Delta. Preliminary Assessment of Socio-economic Impacts and Impacts on Northern Wetlands and Fisheries. Egypt.* Pilot study for a quantitative approach. Delft Hydraulics, Delft, The Netherlands.

EEAA (1995) *Report on Framework of National Action Plan for Dealing with Climate Change.* US Country Studies Program, Egyptian Environmental Affairs Agency (EEAA), Cairo, 62 pp.

EEAA (1999) *The Arab Republic of Egypt: Initial National Communication on Climate Change. Prepared for the United Nations Framework Convention on Climate Change (UNFCCC).*

Egyptian Environmental Affairs Agency (EEAA), Cairo, 160 pp.

Eid, H.M. (1994) Impact of climate change on simulated wheat and maize yields in Egypt. In: Rosenzweig, C. and Iglesias, A. (eds) *Implications of Climate Change for International Agriculture: Crop Modelling Study.* US Environmental Protection Agency, Washington, DC.

Eid, H.M., El-Marsafawy, S.M., Ainer, N.G, Ali, M.A., Shahin, M.M., El-Mowelhi, N.M. and El-Kholi, O. (1996) *Vulnerability and Adaptation to Climate Change in Egyptian Agriculture.* Country Study Report, US Country Studies Program, Washington, DC, pp. 1–53.

Eid, H.M., El-Raey, M. and Frihy, O. (2001) Using MAGICC and SCENGEN climate scenarios generator models in vulnerability and adaptation assessments. Paper presented to Meteorology and Environmental Issues Conference, March 2001, Egypt.

El-Raey, M., Nasr, S., Frihy, O., Desouki, S. and Dewidar, Kh. (1995) Potential impact of accelerated sea level rise on Alexandria governorate, Egypt. *Journal of Coastal Research* (Special Issue) 14, 190–204.

El-Raey, M., Fouda, Y. and Nasr, S. (1997) GIS assessment of the vulnerability of the Rosetta area, Egypt to impacts of sea rise. *Environmental Monitoring and Assessment* 47, 59–77.

El-Raey, M., Frihy, O., Nasr, S.M. and Dewidar, K. (1999) Vulnerability assessment of sea level rise over Port Said Governorate, Egypt. *Environmental Monitoring and Assessment* 56(2), 113–128.

El-Shaer, M.H., Rosenzweig, C., Iglesias, A., Eid, H.M. and Hellil, D. (1997) Impact of climate change on possible scenarios for Egyptian agriculture in the future. *Mitigation and Adaptation Strategies for Global Change* 1, 233–250.

2.6 Ghana

William Kojo Agyemang-Bonsu[10]

2.6.1 Introduction

Ghana signed the United Nations Framework Convention on Climate Change (UNFCCC) in June 1992 and ratified it in September 1995. The UNFCCC recognizes climate change as a major threat to the world's environment and development aspirations. For Ghana, being a tropical and low-lying country and heavily dependent on agriculture for both export and domestic use, the potential impacts of climate change such as sea level rise (SLR), and change in local climate conditions (such as temperature and precipitation) could have important and significant negative impacts on the coastal zone, food production, availability of biomass and hydro-energy sources as well as the level of biodiversity. Climate change could also significantly affect sources of water supply, human health and terrestrial and aquatic ecosystems.

The UNFCCC commits the world at large to deal with the problem of climate change by limiting the emission of greenhouse gases (GHGs) and addressing adaptation should there be climate change. As a Party to the Convention, Ghana is required under Article 4 paragraph 1 and Article 12 of the Convention to prepare a 'National Communication', including the inventories of GHG emissions and their removal by sinks, as well as information on policies and measures taken to adapt to the impacts of or mitigate climate change.

Under the NCCSAP Phase 1, Ghana carried out two important vulnerability and adaptation assessments in the areas of water resources and coastal zones. Two separate reports on water resources and coastal zone vulnerability to climate change have been published. Major highlights from these reports were captured in Ghana's *Initial National Communication* (Environmental Protection Agency, 2001). The vulnerability of agriculture to climate change was addressed under the United Nations Development Programme (UNDP)/Global Environment Facility (GEF) enabling activities. The water resources vulnerability assessment study is addressed here.

Water resources are needed for domestic and industrial water use, sanitation, irrigation of crops, generation of hydropower, navigation, etc. Hence, they are essential for socio-economic developments.

The rapid increase in the country's population will add to the current pressures on the

---

[10] Environmental Protection Agency, PO Box M326, Accra, Ghana. Email: wbonsu@epaghana.org

quantity and quality of water. In general, climate change may put further constraints on the water resources because of changes in spatial and temporal distribution of the resources. The study therefore considered the potential impacts of climate change on flow characteristics of streams and rivers, irrigation water demand, hydropower generation, groundwater recharge and frequency and intensity of floods. In addition, the socio-economic impacts of climate on water resources were assessed.

2.6.2 Approach

A representative basin approach was used because the entire river system in the country could not be assessed due to limited resources. Three basins (Fig. 2.8) were selected representing the Coastal, South-western and Volta basin systems. The assessment was on the impacts of climate change on river discharge, floods, groundwater recharge, domestic and industrial water demand, irrigation water demand, hydropower generation and livelihoods (socio-economic). Teams of experts were constituted as working groups to assess each of these study areas.

Artificial scenarios of climate change were developed for temperature and rainfall. The scenarios were obtained from the base data by uniformly increasing or decreasing the magnitudes of the two hydrometeorological parameters. Scenario runs were carried out with the hydrological/water balance (WATBAL) model to test the sensitivity of the water resources to climate change. General Circulation Models (GCMs) and Regional Atmospheric models, which had been based on scientific principles, were used in deriving the most probable climate change scenarios. The hydrological model was then run with these GCM-based climate change scenarios.

Different time horizons for the climate change analyses were selected. The period of 1961–1990 was selected as the base period, while impact and vulnerability assessments were carried out for 1991–2020 and 2021–2050.

Again the WATBAL model was used to assess the impact of climate change on stream flows. The model was calibrated and then validated over different time periods of the rainfall and temperature data. The calibration of the model was facilitated by the prior analysis of the historical data sets. The impacts on runoffs were assessed by comparing the simulated runoffs of the future climate change scenarios and the base climatic scenario.

Impacts of climate change on groundwater resources were assessed on the basis of groundwater recharge. The outputs from the hydrologic modelling were used in a water balance analysis to compute the changes in recharge under the climate change scenarios. The impacts of climate change on the following water resource systems: irrigation, water supply for domestic and industrial use and hydropower generation, were carried out using the results of the hydrologic modelling in a water resource system model, Water Evaluation and Planning System (WEAP).

The CROPWAT model, which models water requirements for crop growth, was used to compute the potential crop water requirements and thus irrigation water demand. The results were used as input for the WEAP model for the various time horizons up to year 2050 in order to assess irrigation water demand.

The incidence of floods under climate change could not be assessed by any of the models because of the monthly average data used in the studies. Thus a literature study was carried out to find state of the art methods for assessing the potential of flood hazard under climate change. Empirical approaches were attempted in this direction.

Although vulnerability can be associated with any of four factors (annual availability, intertemporal distribution, water quality, and water requirement) most studies have measured vulnerability on the basis of annual water quantity (Kulshreshtha, 1993). Vulnerability is often expressed in one of the following alternative ways: water dependency, water resources constraints, water deficit, demand-supply balance and joint availability and use level criteria. In this study, the joint demand and supply level (JDSL) criteria described in Kulshreshtha (1993) were used. The JDSL criteria are based on both the available supply on a per capita basis and its relative utilization. The latter is defined as the demand–supply

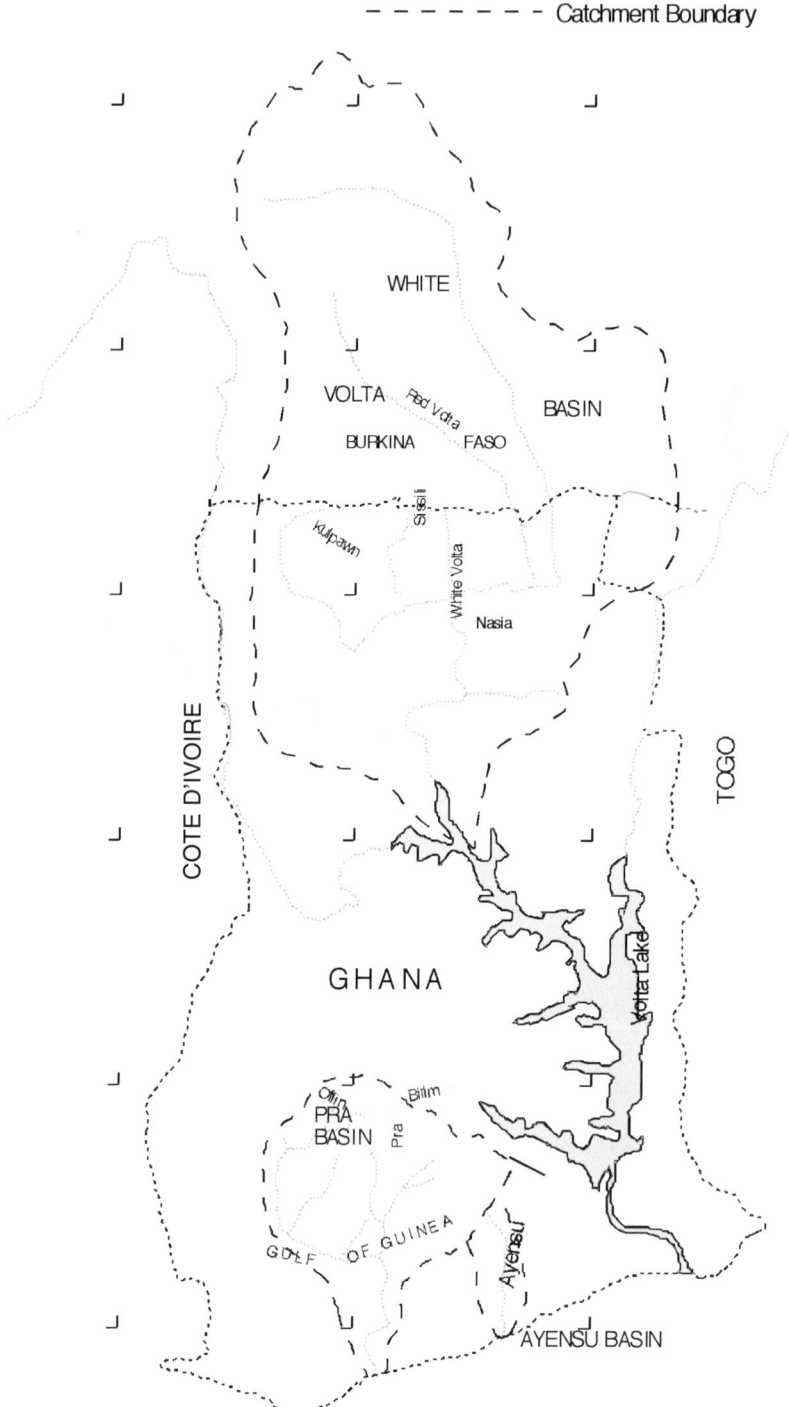

Fig. 2.8. Representative basins.

ratio. After evaluating the vulnerability of the country's water resources to climate change, there was the need to evaluate adaptation measures, which should be put in place to reduce the vulnerability. The adaptation options were assessed based on the approach given in Feenstra et al. (1998) and classification and adaptation strategies development were assessed using the approach given by Carter et al. (1994).

2.6.3 Results of the water resources vulnerability study

The results of the climate change impacts on water resources are discussed in more detail in terms of river discharge, groundwater recharge, irrigation, hydropower generation and socio-economic conditions, in the ensuing paragraphs. Box 2.3 gives a summary of the study results. Impacts of climate change in the areas of floods and domestic and industrial water demand were not assessed in detail. They are discussed briefly below.

Floods are usually associated with rainfall intensities of short duration, a few hours or days. The GCM output and climate change scenarios developed only presented monthly average rainfall data. Relationships between historic 24 h maximum rainfalls and monthly rainfall totals were investigated, but were inconsistent. Hence, impacts on floods could not be assessed. However, available literature

---

Box 2.3. Summary of general findings of the study.

1. There was observed increase in temperatures of about 1°C over a 30-year period and reductions in rainfall and runoff in the historical data sets.
2. Runoffs or discharges in all the representative basins are sensitive to changes in precipitation and temperature and thus to climate change. A 10% change in precipitation or a 1°C rise in temperature could cause a reduction in runoff of not less than 10%.
3. Simulations using climate change scenarios indicated reduction in flows between 15–20% and 30–40% for the years 2020 and 2050, respectively.
4. The last decade (1986–1995) had the most devastating rainfall events and a relatively high number of 24-h maximum rainfall events. Increases in historical temperature and evapotranspiration tend to support the occurrence of this physical phenomenon. A historical analogue may point to increases in flood frequency under climate change.
5. Climate change could cause reduction in groundwater recharge between 5 and 22% by the year 2020. Reductions for the year 2050 are projected to be between 30% and 40%.
6. Domestic and industrial water demand may not be affected by climate change but this needs further assessment.
7. Irrigation water demand could be affected considerably by climate change. In the humid part of the country, the increases in irrigation water demand due to climate change by 2020 and by 2050 are about 40% and 150% of the base period water demand. For the dry interior savannah, the corresponding increases in 2020 and 2050 are about 150% and 1200%, respectively. Further modelling of crop water requirements indicates that vegetables have higher sensitivities to climatic change than cereals.
8. Hydropower generation could be seriously affected by climate change. The projected reduction by 2020 is about 60% from the base value in the Pra basin modelled by WEAP.
9. From a socio-economic point of view, there may be secondary impacts on health, nutrition and energy-based industrial activities if proper adaptation options are not embarked upon.
10. A vulnerability index involving the application of water availability and use criteria indicated that the country had water surplus in the base year (1990), except in the coastal basin. The coastal basin had water management problems in the base year. By the year 2020 and beyond all the basins will be marginally vulnerable. That is, the country will face water management problems.
11. The use–availability ratios in the country are very small. The values are 2–10% and 5–31% in 2020 and 2050, respectively. Should this ratio increase to 40%, to meet food security and for export, the whole country will be vulnerable.
12. Adaptation options suggested were in general for water conservation and efficient use of water for projected reduction in water resources.

(Dankwa, 1974; Mott MacDonald, 1991; Opoku-Ankomah and Amisigo, 1998; Opoku-Ankomah and Forson, 1998) indicated that in parts of the country devastating rainfall events and a relatively high number of 24 h maximum rainfall events occurred in the last decade (1986–1995). Climate change scenarios also indicate an increase in temperature in all basins studied. An increase in air temperature and resulting evapotranspiration is known to result in larger thunderstorms and a greater risk from flash flooding (Opoku-Ankomah and Amisigo, 1998). When assuming climate change is already occurring today, this could indicate that the observed increases in flood frequency in recent times may not be merely a natural variability as stated in the literature. Further studies will be needed in this area.

Domestic water demand is driven, to a large extent, by the population growth and socio-economic development (including industrial development). The direct impact of climate change, in terms of temperature and precipitation changes, on water demand by most industries is envisaged to be very small (i.e. mainly on cooling processes). This was not assessed in more detail. Due to lack of data, the impacts of climate change on domestic water demand could also not be assessed.

River discharge

Impact assessment on stream flow, and thus the surface water resources, was carried out through model simulations using WATBAL. The results indicate that runoff is sensitive to changes in precipitation and temperature in all the representative basins. The magnitude of the changes is approximately similar in all the basins, however, the changes were not uniform. A 10% change in precipitation at constant temperature produced between 10% and 25% changes in runoff depending on the magnitudes of precipitation and temperature at which the changes occurred. Similarly, for a 1°C rise in temperature, there was a reduction of about 10–23% in runoff.

Simulations using GCM-based scenarios indicated reductions in flows for the future. Reductions in flows for the year 2020 are 17%, 20% and 16% for Pra, Ayensu and White Volta basins, respectively, compared to the base period 1961–1990. Similarly, for the year 2050, the reductions are 33%, 37% and 37% for the corresponding basins compared to the base period.

These impacts of climate change on stream flows could have serious consequences on the nation's socio-economic development if proper adaptation strategies are not put in place.

Computed vulnerability indices based on available surface water resources and population indicated a condition of water scarcity (extreme vulnerability) for all the basins apart from the Volta basin for the period 1991–2020. The White Volta basin indicated a water stress (vulnerable) situation. The period 2021–2050, however, showed water scarcity for all the basins.

Groundwater recharge

Groundwater resources are the main sources of domestic water supply in villages and some small towns in Ghana. Thus recharge of groundwater from a supply point of view is important. Impacts of potential climate change on groundwater recharge in the representative basins were evaluated from the subsurface and base flow components generated from the WATBAL model. The recharge was estimated as the sum of subsurface flow and base flow.

In the base period, the mean annual recharges for Pra, Ayensu and White Volta basins were estimated at 3.83, 0.13, and 3.78 km^3/year, respectively. In terms of depth, the corresponding mean annual recharges were estimated at 0.456, 0.207 and 0.224 mm/year. Thus the estimated recharges in the relatively dry Ayensu and White Volta basins are about half of those in the humid Pra basin.

There were reductions in recharges in all the representative basins for the climate change scenarios in the years 2020 and 2050. For the year 2020, the reductions are 17%, 5% and 22% for Pra, Ayensu and the White Volta basins compared to the base period, respectively. The reduction in the Ayensu basin was unduly low. The reductions for the year 2050 were even greater and are 29%, 36% and 40% for Pra, Ayensu and White Volta basins compared to the base period, respectively.

Comparison of the reductions in recharges of the groundwater to reduction in average stream flows showed similar results. This implies that climate change will equally impact average stream flow and groundwater resources.

Irrigation

Irrigation of crops in the country is traditionally on a low scale due to the country's dependence on rainfed agriculture. Irrigation of crops is, however, bound to increase to feed the rapidly rising population.

In assessing the impacts of climate change on irrigation water demand, the CROPWAT model was employed to determine the net irrigation water requirements using the temperature and rainfall scenarios for the base period and for the years 2020 and 2050. The net irrigation water demand was converted to gross water demand by dividing by the local efficiency factor of 0.54.

The gross water demand in the representative basins was determined for the year 2020 based on the areas planned to be put under irrigation by Ghana Irrigation Development Authority (GIDA). These GIDA plans (Ghana Water and Sewage Corporation (GWSC), 1998a) did not consider climate change. For the year 2050, the areas to be irrigated were estimated based on population increase from 2020–2050. The water demand was computed with and without climate change as a factor.

The results in the Pra basin indicate that the water demand for the year 2020 would increase by 510% without climate change and 551% with climate change, from the base period 1961–1990. Thus the change due to climate change alone was estimated at 41% of the base value (4,200,416 m^3). Similarly, for the year 2050 in the Pra basin, the changes with respect to the base values are 771% without climate change and 922% with climate change. Thus the change due to climate change alone in 2050 is 151% of the base value.

In a similar analysis, water demands for the years 2020 and 2050 in the Ayensu basin for climate change alone are 141% and 652% of the base value (48,128 m^3), respectively.

The water demands in the White Volta basin for the years 2020 and 2050 for climate change alone are 278% and 1206% of the base value (6,056,400 m^3), respectively.

It is interesting to note that the changes in area under cultivation from the year 2020 to the year 2050 in all cases were slightly less than 50% while the changes in the water demand due to climate changes differed by about a factor of four. Furthermore, the results of the study showed that vegetables have higher sensitivities to climatic change than cereals, for example, rice (Table 2.5).

Hydropower generation

Hydropower generation is determined by the head of water behind a dam on a river. The head is also related to the amount of water behind the dam. In assessing the impacts of climate change on hydropower generation, the WEAP model was used in routing the flows through the dams for hydropower generation. The WATBAL model was first used in simulating the flows under different climate change scenarios. The output was then expressed as

Table 2.5. Increases (%) in crop water requirements with temperature sensitivity.

Basin	Crop	Year (temperature)			
		2020 (1.5°C)	2020 (2.5°C)	2050 (2.5°C)	2050 (4.5°C)
Pra	Rice	1.9	4.6	12.1	9
	Vegetables	2	10.6	27.1	14.6
Ayensu	Rice	2.1	0.1	3.1	9.1
	Vegetables	2.7	4.2	8.5	13.5
White Volta	Rice	10.9	1.6	6	10.7
	Vegetables	11.9	4.6	13.9	15.7

hydrological fluctuations and used as input for the WEAP model.

The WEAP model was run for the Pra basin only. The inflows showed a reduction of about 45%. The energy generated for the base case (1990) was about 108 and 160 GWh for Awisam and Hemang, respectively. However, for the altered climate scenarios, the energy generated was about 41% of the base values, a 59% reduction for the year 2020.

Socio-economic conditions

Only second-order climate change impacts on livelihoods (socio-economic conditions) due to changes in supply and demand of the water resource were analysed. The findings indicate that both surface and groundwater resources will decrease across the country in the years 2020 and 2050. Water demand will, in general, increase due to an increase in population and the need for improvement in socio-economic conditions. Moreover, as indicated above, climate change will increase irrigation water demand in the country.

Simulation of hydropower, which formed 77% of electric energy at the time of assessment, indicated considerable reduction in hydropower output in the order of 59%. Further, the detailed vulnerability index, which was a less pessimistic indicator, showed marginal vulnerability for the years 2020 and 2050 (Table 2.6).

From the foregoing, there could be impacts on health, nutrition, employment, energy-based industrial activities, etc., if proper adaptation options are not embarked upon. Consequently two categories of adaptation options were proposed to address water demand and supply side management.

Adaptation options to be embarked upon from the supply point of view include:

- inter-basin water transfer, e.g. from the Volta basin to the coastal basin;
- changing location or height of water intake points using floating intake structures;
- installing canal linings and using closing conduits instead of open channels in transporting water, to say, irrigation fields;
- integration of separate reservoirs into an integrated system;
- artificial recharge of groundwater to reduce evaporation;
- building of reservoirs on rivers, which have run-of-the-river intake points;
- good land use practices to discourage siltation of dams and thus maintain live storage for hydropower generation, domestic and industrial water supply, especially during dry seasons.

Adaptation from demand points of view include:

- reduction in transmission losses of water to demand sites;
- efficient domestic appliances;
- dual supply system, potable and non-potable;
- recycle domestic water for non-potable uses;
- reduction in irrigation water use, e.g. night-time irrigation, drainage reuse and use of water effluent;
- introduce low water use crops;
- introduce high value per water use crops;

Table 2.6. Vulnerability index (persons per million m^3 of surface water).

Basin	Year		
	1990	2020	2050
Pra	442[b]	1229[d]	3141[d]
Ayensu	948[c]	3064[d]	7768[d]
White Volta	301[b]	927[c]	2468[d]
National	354[b]	1313[d]	2789[d]

[a]Water surplus (not vulnerable); [b]Marginally vulnerable (water management problems); [c]Vulnerable (water stress); [d]Extremely vulnerable (water scarcity).

- change irrigation systems to drip, micro spray;
- reduction in industrial water use through reuse of acceptable quality water, recycling and dry cleaning technologies;
- efficient water use for energy generation, e.g. keeping reservoirs at lower head to reduce evaporation;
- build additional reservoirs and hydropower stations;
- construct low head run-of-the-river hydropower;
- introduce more efficient hydropower turbines.

2.6.4 Experiences and lessons learned

It has been realized that although the independent sectoral vulnerability studies lead to in-depth information and important conclusions, our experience now shows that we could have achieved better results and comprehensive policy options and above all, the country would have derived much more benefits, if various working groups for the vulnerability and adaptation studies had talked to each other during the period of assessment. This supports the point that integrated vulnerability and adaptation assessment approaches are more profitable to a country than any detailed independent sectoral studies.

The greatest challenge during the studies was the availability of relevant data, especially for domestic and industrial water demand. The combined demand–supply criterion considers water use explicitly and the low ratio of water use to supply in the country gives it a less pessimistic picture.

2.6.5 Follow-up research

The following follow-up studies are proposed:

- There are strong seasonal variations in rainfall and runoff in the country; however, the WATBAL model did not have seasonal parameters to take that into consideration. Hydrological modelling with seasonality aspects will be relevant to undertake. Impact assessments based on seasonal values will give a better picture of vulnerability in the basins.
- Changes in inter-annual variability of rainfall and runoff under climate change could pose a more serious threat to human survival than changes in the long-term means. This aspect of the study will need consideration.
- Climate change, climate variability and land use can jointly impact water resources. There will be the need to model these factors in our tropical environment.
- On the recharge studies, a direct modelling approach will be needed for more accurate results. There will be the need to measure some field parameters for the modelling.
- Further investigation will need to be undertaken to assess the impact of climate change in high intensity rainfall events and flood frequency.
- In this study, it was assumed that the impacts of climate change on domestic and industrial water demands are very small. Existing literature on the issue indicates a relationship between rainfall and domestic water demand. This will need to be investigated.
- Further work will be needed to estimate irrigation water demand for crop production to meet food security under climate change.
- There will be the need to undertake river basin management using WEAP in an integrated manner to resolve potential conflicts. The Volta basin should be studied in an international context. Conflict between irrigation development and hydropower generation and other uses will be examined.

Furthermore, a regional programme to follow-up on the outcomes of these studies for the management of the Volta basin should be implemented.

The follow-up activities have been presented to the Global Environment Facility and NCCSAP Phase 2 for funding. In the NCCSAP Phase 2 we anticipate assessing the cumulative impact of climate change and climate variability on land use and water resources.

2.6.6 Policy implications

The methodologies and studies were in line with policy needs, consequently a number of policy initiatives have been implemented that either fully or partially respond to the outcomes of the studies.

As a result of the reduced power output from the Akosombo Dam the government of Ghana is now actively encouraging investors who wish to establish electric power generation to invest in renewables, especially wind and solar power. Various projects are underway for the construction of solar-powered generation. Other projects considered are a pilot project on the construction of wind-powered generators. In this project the feasibility of wind energy in Ghana should be determined before wider scale implementation.

The government has restructured the power sector with the view of improving the supply situation and introducing competition into the power generation sector by encouraging independent power producers. This is to ensure better efficiency in the power sector.

Several years of reduced hydropower supply owing to reduced inflows and increasing energy demand because of increasing population and economic activities, have led to a number of demand side management problems.

Capacitor banks in high-energy consuming industries have been installed in order to correct power factors to reduce losses in the power system. Nationwide campaigns have been undertaken in order to educate the public in ways of reducing their energy consumption. This was done by publicity through newspapers, radio and television advertisements and announcements. Seminars were also organized countrywide to educate the public on the hydropower generation situation and to urge them to use electric power judiciously. The Energy Foundation has developed a chart on relative power consumption of power appliances and has prepared and published energy efficiency standards and labels. This is to enable the general public to be aware of which of their appliances consume more energy so that they can manage their consumption better. Additionally, Ghana has constructed two thermal plants, which deliver a total of 300 MW of power.

The Volta River Authority, currently the only hydropower generator, imported power and is promoting the sale of compact fluorescent lamps (CFL), which have a higher efficiency than ordinary incandescent bulbs. The sales are being promoted through advertisements in the newspapers, radio and television. To encourage the use of the CFLs, government has removed all taxes (including import duty and value added tax resulting in a 35% reduction of the wholesale price).

The concept of demand side management is important, there is the need to install or build very efficient infrastructure capable of maintaining the right water pressures and water heads. In cases where there may not be adequate surface water, the infrastructure should either be improved to handle excess water from inter-basin transfers or supplemented with groundwater resources. There should also be proper conservation of water during the wet season for usage in the dry season due to the spatial and temporal nature of the rainfall pattern in Ghana.

To promote efficient use of water, Ghana Water Company has been allowed to charge economic rates to sustain its systems and also to serve as a deterrent to consumers on water waste.

Additionally, there has been massive public education of farmers along riverbanks. Government as well as civil society have embarked on reforestation projects along the banks of some rivers.

Currently the government is considering the promotion of rain harvesting in the northern section of the country to ensure that water is made available for farming and other uses. The major challenge of the studies has been how to integrate and mainstream their outcome into the national development agenda (i.e. the Ghana Poverty Reduction Strategy (GPRS)). As part of the ongoing strategic environmental assessment by the Environmental Protection Agency on the GPRS, a window of opportunity has been created that allows for the mainstreaming of climate change concerns, including adaptation, into the GPRS. Full integration of climate change adaptation is, however, not envisaged until a national adaptation plan of action has been developed. It is the government's plan to develop the adaptation plan of action as part of the NCCSAP Phase 2.

The focus of future activities should be to

formulate climate change policies that are consistent with the national poverty reduction strategy, and thus facilitate the mainstreaming of these policies into district as well as national development plans.

This could be achieved through:

1. Identification (through investigations) of the biophysical and socio-cultural-economic environments, a range of methods by which Ghana may improve its capacity to respond effectively, efficiently and sustainably to future climate change.
2. Incorporation of experience gained from the critical examination of present climate-related problems in the design of appropriate responses to anticipated climate change.
3. Identification of transferable models of good practice in response to present climate-related hazards.
4. Involvement of a wide range of stakeholders from governmental institutions and civil society in these investigations, in the knowledge that appropriate responses will involve the whole of society.
5. Examination of the plight of the most vulnerable members of society in relation to climate-related threats, with particular reference to livelihood strategies.
6. Raising awareness of the threat of climate change among policymakers, the private sector and civil society.

2.6.7 Conclusions

The results from the water resources studies have proven to be invaluable as they give a clear indication that the adverse impacts of climate change on Ghana's water are real. A country that once depended solely on hydropower now has to resort to thermal generation to meet increasing demand.

The studies have revealed that the impact of climate change on water resources has a number of consequences – outlined above – on the well-being of the people of Ghana, and efforts by government have been taken to address these impacts. However, the major challenge that faces the country is the effective implementation of the recommendations from the studies. It has become increasingly apparent that there is a need to develop a comprehensive policy for climate change adaptation that also addresses issues of land management and vulnerability of the poor with particular emphasis on gender. Therefore the focus of future programmes in the area of adaptation, including work on water resource assessment, should be on the examination of the linkages between poverty and climate change and the consequences of climate change on the livelihood systems of poor communities.

The water dependency criterion indicated water scarcity or extreme vulnerability. The water dependency method is less detailed than those investigating the consequences of climate change on the livelihood systems of poor communities and these may be considered as more appropriate.

Hydropower generation could be judged to be extremely vulnerable by 2020 owing to over-dependence on this cheap source of electricity in the country. The reduction in power generation in the order of 50% simulated for the Pra basin in 2020 underlines the fact that existing supply cannot meet future demand. Again, following from the results of the study it was recognized that the education of the general public on the economic use of water is important.

It was also shown that there is a need for crop diversification by shifting to more drought resistant crops that require less water. Other measures include the improvement of irrigation techniques and practices through the adoption of those that use water more efficiently, e.g. unlined canals should be lined to prevent seepage and open main canals will have to give way to pipes. Moreover, efficient irrigation methods like drip irrigation should be preferred to less efficient ones like flooding and furrow and mulching, and other water conservation methods will have to be applied in the fields. Finally, storage capacity (reservoirs) for irrigation must be increased wherever possible instead of allowing rivers and streams to flow into the sea unused.

Main Publications From the Research

Environmental Protection Agency (2001) *Initial National Communication of*

Ghana. Environmental Protection Agency, Accra, Ghana, 172 pp. Available from: http://unfccc.int

Opoku-Ankomah, Y. (1998) *Volta Basin System Surface Water Resources in Water Resources Management Study. Information Building Block. Part II, Vol. 2.* Ministry of Works and Housing, Accra, Ghana.

Opoku-Ankomah, Y. and Amisigo, B.A. (1998) Rainfall and runoff variability in the south-western river system of Ghana. In: Servat, E., Hughes, D., Fritsch, J.M. and Hulme, M. (eds) *Water Resources Variability in Africa during the XXth Century.* Proceedings of the Abidjan '98 Conference, Cote d'Ivoire, November, 1998. IAHS (International Association of Hydrological Sciences) Press No. 252, Wallingford, UK, 462 pp.

Opoku-Ankomah, Y. and Forson, M.A. (1998) Assessing surface water resources of the South-Western and Coastal basin systems of Ghana. *Hydrological Sciences Journal* 43, 733–740.

2.7 Kazakhstan

Irina Yesserkepova, Boris Stepanov, Vsevolod Golubtsov, Valery Lee and Roza Yafyazova[11]

2.7.1 Introduction

Kazakhstan is situated in the central part of the Eurasian continent and has a dry continental climate. Preliminary impact and adaptation studies have shown that the natural resources and the economy of Kazakhstan are vulnerable to climate change. The government of Kazakhstan supports climate change research activities including scientific research as a background for environmental protection policy actions and climate change strategy development for the future.

After signing and ratification of the United Nations Framework Convention on Climate Change (UNFCCC), Kazakhstan started conducting climate change study projects and related activities with the support of the US Country Studies Program in 1995. During this project the first greenhouse gas (GHG) inventory, vulnerability and adaptation assessments and mitigation analysis were carried out. Due to the large territory of the country, the diversity of climate conditions and the large variety of economic activities, it was not possible to cover all vulnerable sectors and regions in one study. It was required to focus the climate studies on the regions that are very important for the economic development of Kazakhstan. The densely populated foothills and mountain regions in the south-eastern part of the country and the oil and gas-extracting regions in the Caspian Sea coastal zone urgently needed a new strategy of prevention measures against mudflows and snow avalanches, and storm surges, respectively.

Climate change is expected to increase the risk of mudflows and snow avalanches pressing the need for a vulnerability and adaptation study. Furthermore, the Initial National Communication of Kazakhstan to the UNFCCC had not been prepared and a GHG inventory was required in accordance with the revised international methodology. The NCCSAP activities contributed to the continuation of these studies.

This part of the country section describes the results of the climate change impact, vulnerability and adaptation studies for Kazakhstan's part of the Caspian Sea coastal zone, as well as snow avalanches and mudflows in the mountains of south-east Kazakhstan. The reason for selecting these studies to be presented in this section in more detail is that there were only a few publications on scientific results in Kazakhstan or abroad. This book gives us an opportunity to distribute the results of these investigations.

The goals of the study were:

- to develop baseline and climate change scenarios for the studied regions;
- to conduct vulnerability and adaptation assessments of the Caspian Sea coastal zone and mountain and foothill areas under possible climate change;

---

[11] Kazakh Research Institute for Environment Monitoring and Climate (KazNIIMOSK), pr. Seifullina, 597, 480072 Almaty, Kazakhstan. Tel.: 3272 676487, Fax: 3272 542285, Email: irina@kniimosk.almaty.kz

- to prepare the recommendations to the government of Kazakhstan in order to protect its coastal zone and for preventing possible damage induced by mudflows and snow avalanches;
- to conduct cost evaluation of adaptation measures;
- to prepare recommendations to the government for possible adaptation measures.

2.7.2 Approach

The approach of the study was based on the Intergovernmental Panel on Climate Change (IPCC) methods of assessing the impacts from and potential adaptation to climate change (Carter et al., 1994). This approach follows a seven step strategy according to IPCC technical guidelines for climate impact and adaptation assessment (see also Section 1.5):

1. Definition of the problem.
2. Selection of the method.
3. Testing of the method.
4. Selection of scenarios.
5. Assessment of impacts
6. Assessment of autonomous adjustments.
7. Evaluation of appropriate adaptation strategies.

This approach allows different scenarios, scales (global, local) and long- or short-term effects to be taken into account. The retrospective analysis of mudflow activity during the Pleistocene provides a scientific background for the scenarios of change of mudflow activity in the course of global climate warming in the 21st century. The study of avalanche activity was based on the data of observations and climate change scenarios prepared for the region of the study. The Kazakh Research Institute for Environment Monitoring and Climate (KazNIIMOSK) has developed a water balance model for the calculation of seawater levels. The model allows determination of the sea level on the basis of river water inflow, water consumption changes in the basin, evaporation, precipitation and seawater runoff to the Kara Bogaz Gol gulf. Furthermore, a hydrodynamic model for simulation of storm surges developed by the Danish Hydrological Institute (Module MIKE 21) was adapted to the Caspian Sea conditions.

2.7.3 Results of the vulnerability assessment

The results of the vulnerability to climate change impacts and adaptation options for Kazakhstan's part of the Caspian Sea coastal zone and mountain region of south and south-east Kazakhstan are described here.

Climate change scenarios

THE CASPIAN SEA REGION. The general atmosphere circulation models were used to define potential changes in the values of temperature, precipitation and evaporation for the Caspian Sea area under climate change. The models predict a mean annual air temperature increase in the Caspian Sea area of 3.7–4.9°C, while the annual precipitation will increase by an average of 52 mm (GFDL, CCC and UKMO models) or decrease by 4–8 mm (GISS model). The effective evaporation (evaporation minus precipitation) may increase to 158–243 mm under the changed climate.

SOUTH-EAST KAZAKHSTAN. In accordance with the existing climate change scenarios in south-east Kazakhstan in the coming 25–30 years, climate change can lead to increase of air temperature by 2–3°C (KazNIIMOSK, 2000). To assess the effects of climate change a number of scenarios were constructed. Currently, no reliable prediction can be made for regional climate changes, especially in mountainous regions (IPCC, 1996). Only more or less plausible estimations are available, based on General Circulation Model (GCM) outputs and observed tendencies of the current climate. For the mountain regions of south-east Kazakhstan all GCMs suggest an increase in annual temperatures ranging from 3.7°C to 7.1°C under the assumptions of doubling CO_2 concentration, which is expected to occur between 2050 and 2075. In addition, we used incremental scenarios and some statistical relationships between climate characteristics and snow avalanche activity.

The Caspian Sea

IMPACTS. At present the Caspian Sea is a closed reservoir, its surface is almost 27 m below the World Ocean Level. Results of calculations of changes in runoff and effective evaporation in the Caspian Sea basin under climate change were used for modelling changes of the sea level for the first half of the 21st century. According to the data of the Russian State Hydrological Institute (SHI), the water consumption in the Caspian Sea basin reached 40 km^3/year by the end of the 1980s. In the mid-1990s the water consumption decreased to 30 km^3/year as a result of the general decrease in economic activity. According to SHI this level of water consumption will be maintained till the year 2000, after which a gradual increase will take place to 40 km^3/year in the year 2020, the same level as in the 1980s.

The results of the different models show that climate change in combination with an annual water consumption of 40 km^3 may cause the Caspian Sea level to rise 4.7 m with an 0.1% frequency (once a millennium) according to the CCC model, 6.4 m according to the UKMO model, and 1.0 m according to the GFDL model. Minimum runoff values were obtained using the GFDL model, while the UKMO model gave maximum runoff values.

It was calculated that the level of the Caspian Sea may increase to 5 m relative to its present value (−27 m) as a result of climate change in the middle of the 21st century. In this condition, the probability of the Caspian Sea level rising up to the −25 m level, taking into account the forecasted water consumption in the Caspian Sea basin, will gradually increase. This process will increase the probability of catastrophic flooding of the Caspian Sea coastal zone in the Pricaspian countries in the middle of the next century.

The expected Caspian Sea level rise in combination with storm surges (or wind set-up) will cause high seawater levels. An evaluation of this effect was conducted using a hydrodynamic model for simulation of storm surges, namely the Danish Module MIKE 21 adapted to the Caspian Sea conditions by KazNIIMOSK. This simulation allocated specific surge heights to various zones along the Caspian Sea, and evaluated the subsequent coastal flooding for three values of the background sea level: −27.0 m, −25.0 m and −22.0 m. It was found that the duration, height and intensity of the wind set-up will be increased by 18–20% by the Caspian Sea level rise.

Moreover, during heavy storm surges the coastal zone will be flooded over a width of 30–50 km, because of the low level of the Pricaspian Lowland. The eastern coast of the northern Caspian including the Lower Emba River is the most vulnerable zone. The wind set-up height in this region will increase to 3 m and the seawater will penetrate into the coastal zone over a distance of 50 km. This catastrophic flooding of considerable areas will also lead to the leaching of large amounts of polluting substances into the Caspian Sea.

The expected sea level rise (SLR) will cause an increase of the groundwater level in the coastal zones. This influence was evaluated by the Institute of Hydrogeology and Hydrophysics. Their assessment showed that an increase of the groundwater level will take place in the coastal zone over a width of 5–10 km. The groundwater level will increase in general by 0.3–0.5 m; only near the coastline the increase will amount to 1–2 m, and in a few places to 3–4 m.

The gradual increase of the Caspian Sea level since 1978 has brought considerable damage to Kazakhstan's economic activities at the seashore. When the sea level reached the −27 m level the damage amounted to US$1.1 billion. Due to the expected climate change in the next century the sea level will increase to higher levels. According to the existing climate change models, if the CO_2 concentration in the atmosphere is doubled, the sea level may reach the −22 m level and higher by 2050. If the sea attains the −26 m, −25 m and −22 m levels, the damages would amount to US$2.8, 11.9 and 14.2 billion, respectively. The existing protective structures are insufficient to prevent flooding of former safe areas, settlements and economically important facilities.

ADAPTATION MEASURES. The strategy for the protection of Kazakhstan's Caspian seashore includes: construction of sea defences (dykes), relocation of settlements, protection of oil and gas exploration facilities and infrastructure, protection of agriculture and fishery resources, protection of utilities. Thus, the recommended general plan for protection of the coastal areas from flooding consists of a broad range of adaptation measures. Its basic approach is the construction of frontal and ring dykes whose length will be increased from 330 to 1206 km. These dykes will protect industrial enterprises, main oil and gas fields, 92 settlements, 270,000 ha of agricultural land, and transport and social infrastructure. Protection of the population and the economic infrastructure threatened by flooding in case of a SLR to –26 m is an absolute necessity. This first stage protection will necessitate enlarging the existing sea defences to a total length of 1148 km.

Mudflows

A mudflow is a flow of water so heavily charged with earth and debris that the flowing mass is thick or viscous. At the origin sites of rainfall-caused mudflows the precipitation normally falls as hail and snow. One to two times a century precipitation falls with an intensity that is sufficient to trigger large mudflows that are capable of reaching a debris cone. The last activation of glacial-caused mudflows, caused by disastrous emptying of reservoirs of moraine-glacial complexes, took place in the latter half of the 20th century, i.e. about 100 years after the end of the Small Ice Age (1850).

Former investigations, for example of the northern mountain ranges of the Tien Shan (the Zailiysky and Kungey Alatau Mountains), revealed that the mudflow activity was almost equal to zero under the climate conditions of the Ice Ages and it peaked during the Riss-Wurm interglacial period (120–130 thousand years ago).

During interglacial periods mudflows transport sediments, which accumulate in the mountain valleys. These sediments were reworked to moraines by glaciers during glacial periods. The study of the structure of debris cones of the rivers of the northern slope of the Zailiysky Alatau Mountains, where the towns of Almaty, Kaskelen, Talgar and Issyk (total population of about 2 million people) are located, showed that 90–95% of the volume of mudflow deposits were formed in the Quaternary period. The volume of debris cones varies from 0.8 to 18 km^3, half of their volume consisting of mudflow deposits of the Riss-Wurm interglacial period, when in the Antarctic air temperature was 2.6°C higher than the present day (Stepanov and Yafyazova, 1995).

Construction of the main mudflow-check dams at the rivers of the northern slope of the Zailiysky Alatau Mountains was realized in the period 1960–1990, when the patterns of transport of sediments to the foothill plain (debris cones) had not yet been discovered. Their discovery revealed a discordance between the existing capacities of mudflow-storage reservoirs and the necessary capacity. Due to potential climate change mudflow activity will increase tens to hundreds of times compared with the mudflow activity of the 20th century (Stepanov and Yafyazova, 2001; Yafyazova 2003). At present, the existing defence strategy against mudflows is based on detention of mudflows in the mountain valleys by means of dams. In the future this will not only be economically ineffective, but even dangerous. The overfilling mudflow-storage reservoirs may become the objects of increased mudflow hazard in case of dam failure. Figure 2.9 shows a mudflow in Zailiysky Alatau.

Snow avalanches

In Glasovskaya and Troshkina (1998) possible changes of snow and avalanche characteristics, such as snow depth, number of days with intensive snowfalls (>10 cm/day) and the duration of the avalanche-hazardous period, were assessed for the mountains of the former Soviet Union under CO_2 doubling in the atmosphere. For the development of their climate change scenario they used GFDL model outputs. According to their results the snow depth in the mountains of Kazakhstan and Kyrgyzstan can possibly increase by 0 to 10 cm. At the same time the increase in the number of days with intensive snowfalls is expected to be 5–25%. The duration of the avalanche-hazardous period will be reduced by 10–20 days. In other words, the winters will get shorter, but snowier. It is crucial, however,

Fig. 2.9. A frontal wave of the mudflow of 15 July 1973, enveloped in a mud cloud, entering the reservoir in the Medeo trough (the Tien Shan). Its discharge was 10,000 m^3/s, depth 15 m, velocity 10 m/s and density 2380 kg/m^3.

to know the lower boundary (altitude) above which these trends will take place. The results of Glasovskaya and Troshkina are aggregated for the areas considered. But they are not in contradiction with our results although we have not used the GFDL scenario, as it does not seem to work well for south-east Kazakhstan.

In the near future (10–20 years) no dramatic changes in the snow and avalanche conditions are to be expected in the mountain regions of south-east Kazakhstan. However, as a result of the predicted temperature increase, the snowline is likely to rise and this means that the lower boundary of the avalanche risk zone will also rise.

Under doubling CO_2 conditions, expected between 2050 and 2075, the snowline may rise by about 700 m. That means that the snow conditions of today at 2300 m above sea level (asl) (e.g. at the Shymbulak ski winter resort) will in future prevail at 3000 m asl (e.g. close to the Shymbulak snow avalanche station). In the upper part of the Shymbulak ski area only 180 days of snow coverage (instead of 240) can be expected and the annual maximum snow depth may be reduced to only 75 cm (instead of about 125 cm).

Especially in mid-winter (December–February) more snow and, consequently, more avalanches can be expected down to low or at least medium altitudes. In spring (March–April) snowfall will increasingly turn into rain, with the effect that the snow cover at low to medium altitudes will melt earlier and snow will remain only at high altitudes.

Since in the long term the lower boundary of the avalanche risk zone (today at about 1500 m asl) is likely to rise, low-lying infrastructure, such as the Medeo skating rink (1600 m asl) or the Medeo–Shymbulak road, may be less endangered or at least the period of potential avalanche hazard will be reduced. However, higher up in the mountains the avalanche hazard may even increase owing to the possibly snowier conditions and it may remain a considerable risk, particularly for skiers and mountaineers.

2.7.4 Experiences and lessons learned

The impact and adaptation study used the IPCC climate impact assessment methodology, based on the Seven Steps Method. The obtained results support the government in developing new strategies against mudflows and snow avalanches as well as for protecting the Caspian Sea coastal zone. Experience in the application of climate change scenarios and different physical models was gained. A lot of observational data on climate from meteorological stations and field observations were used. Interesting results were obtained using a multidisciplinary approach with the expertise of specialists from different areas. A comprehensive analysis was carried out for the Caspian Sea coastal zone including effects of storm surges and climate change impacts on underground waters. Policymakers are considering the proposed measures, which indicates that the study was relevant.

2.7.5 Follow-up research

Research on mudflows shows that the mudflow deposits of the Holocene (the last 12,000 years) make only 1–2% of the total volume of mudflow deposits of the Riss-Wurm interglacial period, though in a climatic optimum of the Holocene air temperature exceeded the current air temperature by about 1°C. This leads to the following question: why did the increase of air temperature by 1°C in comparison to the air temperature of the 20th century not lead to a significant activation of mudflows? Unfortunately financial resources to research the answer to this question were lacking.

The snow avalanche study was followed by the introduction of a system to forecast snow avalanches, based on the Swiss model NXD2000. This may provide the avalanche forecaster with information about former days with similar conditions and makes the forecasts more precise. The NXD2000 is in operation for forecasting snow avalanches in Zailiysky Alatau.

2.7.6 Policy implications

The studies described above have indicated a list of measures that would contribute to sustainable development of the mountain and foothill regions of Kazakhstan as well as Kazakhstan's part of the Caspian Sea coastal zone. However, these measures will not be fully realized. The main reasons are the lack of a sense of urgency by the officials (global climate warming will not happen in the near future), and the absence of legislation to regulate the realization of adaptation strategies.

For activation of preventive measures against mudflows and snow avalanches it is necessary to introduce new concepts for defence strategies, to encourage the acknowledgement of the impacts of climate change by the Kazakh government, to develop performance requirements for preventive measures, and to create awareness about the relevance of preventive measures among the population.

2.7.7 Conclusions

The study performed under NCCSAP was very important for the government and demonstrated the necessity of further action to protect natural resources, vulnerable regions and the economy against climate change. New strategies were proposed which are very important even if climate change would not have the expected impact. The strategies are even useful for the current situation.

It should be noted that the conclusions about the possible increase of the Caspian Sea level during the first part of the 21st century are preliminary. Further studies are required. They should be aimed at the use of improved GCMs and the assessment of runoff change of the Ural River and other rivers on the Caspian Sea's western coast (Kura, Terek etc.). Furthermore, it is necessary to obtain precise data on the time of transition of the Caspian Sea level from current climate to the conditions of anthropogenic climate change caused by the CO_2 concentration doubling in the atmosphere. Also it is important to make a detailed assessment of storm surge characteristics and increase of groundwater level in the coastal zone. All these studies will allow us to clarify the adaptation measures of the coastal zone to the possible disastrous SLR and to assess the required expenditures.

The study on formation of mudflows in the northern mountain ranges of the Tien Shan (the Zailiysky and Kungey Alatau Mountains) has shown that the climate is a main factor that defines the mudflow activity in this region. The research in the framework of

NCCSAP was successful because of the integrated approach of the programme: it included climate change – vulnerability – adaptation. As a result of limited financial resources the research of mudflows was confined to the development of evaluation methods of mudflow risk. The study of climate change influence on mudflow activity will allow optimization of measures for damage reduction.

Climate change would, in principle, not limit the development of the Zailiysky Alatau as an important tourist and resort area. Rather, the main limiting factors are found in the legislative, social and cultural sphere. The development and implementation of defence measures against snow avalanches is very important for further sustainable development of the region. Therefore the demand for accurate avalanche forecasts and avalanche warning will rise. Hence, one of the principal measures in developing a defence strategy against snow avalanches is the improvement of the observation system and the continuation of research on snow avalanches and other natural hazards. It is necessary to expand the area of snow cover and avalanche studies to the high mountain zone, and to neighbouring valleys of the Little Almatinka River basin, taking into account the perspective of winter tourism development. The existing defence system is not sufficient to prevent damage from avalanches. Therefore, adaptation strategies include both short- and long-term measures.

Main Publications From the Research

Glasovskaya, T.G. and Troshkina, E.S. (1998) Global climate change impact on the avalanche regime on the territory of the former Soviet Union. In: *Materials of Glaciological Studies. Chronicle Discussions*, Moscow 84, 88–92. (in Russian)

KazNIIMOSK (2000) *Assessment of Impact and Adaptation to Climate Change for Kazakhstan's Part of the Caspian Sea Coastal Sector and Mountain Region of South and South-east Kazakhstan. Summary for policymakers.* KazNIIMOSK, Almaty, Kazakhstan, 51 pp.

Stepanov, B.S. and Yafyazova, R.K. (1995) Features of formation of debris cones on the northern slope of the Zailiysky Alatau Mountains. *Hydrometeorology and Ecology* 3, 18–28. (in Russian)

Stepanov, B.S. and Yafyazova, R.K. (2001) Radical review of the defence strategy against mudflows is a necessary condition for the sustainable development of the mountain and piedmont regions of Kazakhstan. In: KazNIIMOSK (ed.) *Problems of Hydrometeorology and Ecology.* Proceedings of the International Scientific and Practical Conference, 12–13 September 2001, Almaty, Kazakhstan, pp 32–35. (in Russian)

Yafyazova, R.K. (2003) Influence of climate change on mudflow activity on the northern slope of the Zailiysky Alatau Mountains, Kazakhstan. In: Rickenmann, D. and Chen, C. (eds) *Mechanics, Prediction and Assessment.* Proceedings of the Third International Conference on Debris-Flow Hazards Mitigation, 10–12 September 2003, Davos, Switzerland, pp. 199–204.

2.8 Mali

Abdoulaye Bayoko[12]

2.8.1 Introduction

Mali is a Sahelian country situated in West Africa. Its economy is essentially based on agriculture and livestock farming, which are both strongly dependent on the availability of water resources.

In the agricultural sector cotton is an important export product bringing hard revenues, while maize is an important food crop for the entire population of the country. Livestock is the third largest export product of Mali after cotton and gold. Livestock-based income was valued at US$287 million in 1998, equal to 9.28% of the gross domestic

---

[12] National Centre of Scientific and Technological Research, BP 3052 Bamako, Mali, Email: projetgef@afribone.net.ml

product (GDP). Exports of livestock during the same year were US$122 million.

Considering the importance of agriculture (i.e. maize and cotton production) and water resources to the national economy, these sectors were selected for an assessment of climate change impacts, vulnerability and adaptation under the NCCSAP. The studies were based on earlier studies performed under the US Country Studies Program among others, which were also implemented under the United Nations Framework Convention on Climate Change (UNFCCC) in Mali. The studies enabled the expert teams in Mali to improve methodologies used in the previous studies and refine results using more plausible climate and socio-economic scenarios for Mali (CNRST, 2003a).

Two basins were selected for the assessment: Baoulé to Bougouni and Sankarani to Sélingué, as indicated in Fig. 2.10.

2.8.2 Approach

The studies were carried out using the basic climate change vulnerability and adaptation assessment approach described in Carter et al. (1994) and Feenstra et al. (1998).

First meteorological (rainfall, temperature,

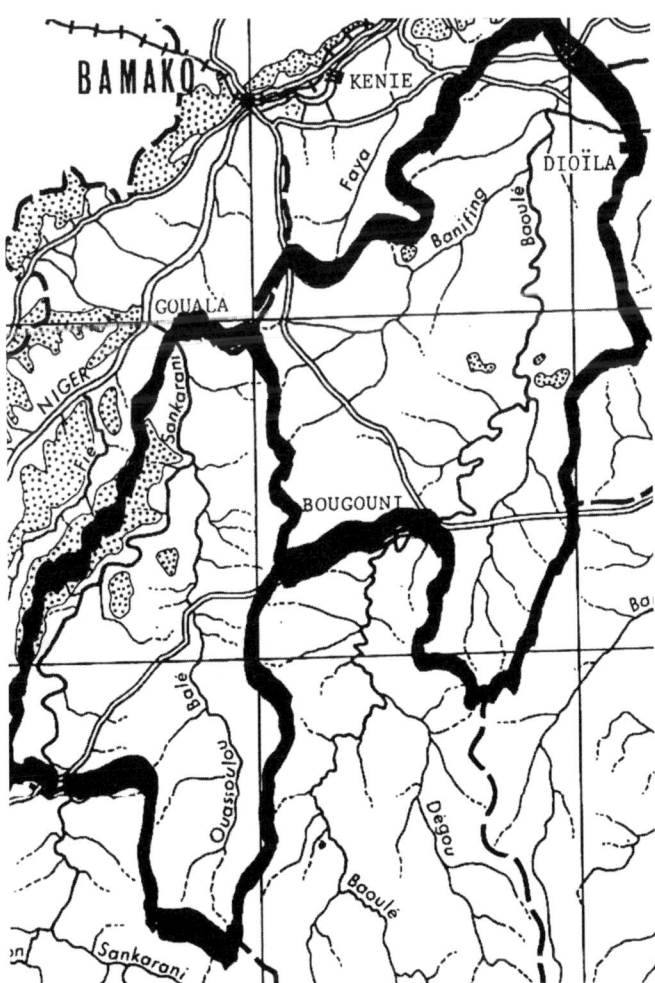

Fig. 2.10. The area of the study includes the two basins of the Sankarani (left area) and of the Baoulé.

etc.), socio-economic activity and hydrological data were collected. To enable climate change scenario development, historical meteorological data were also collected for other locations in Mali. Additionally socio-economic data were collected for the country as a whole. Then climate change and socio-economic scenarios for the years 2000 (the base year), 2025, 2050, 2075 and 2100 were developed using the methodologies described in Feenstra et al. (1998) and Carter et al. (1994) and MAGICC/SCHENGEN described in Wigley (2004).

This was followed by an analysis of the hydrological parameters in the study areas (including among others groundwater levels and the available water volumes etc.) for the base year and the identified future years. After assessing the water needs for the different socio-economic activities (agriculture, livestock, electricity production and fisheries), a comparison was made between water supply (surface and groundwater) and demand for the socio-economic activities. This gave the supply–demand ratio and identified the possible shortfalls in water supply for the two study areas.

The impacts of climate change on water resources were assessed using the following models:

- RAINRU (Professor H. Savenije, Institute for Water Education (IHE) of Delft, The Netherlands) for the determination of runoff and the available water volumes in the basins according to rainfall;
- CRIWAR (Crop Irrigation Water Requirement) (International Institute for Land Research and Improvement, Wageningen, The Netherlands) for the determination of water needs by plants according to the meteorological conditions; and
- an empirical model (Doorenbos and Pruitt, 1986) to translate the impacts of deficits in water on cereal yields.

The model results combined with scenarios of water demand for domestic and industrial use and for other socio-economic activities, gave the estimated socio-economic impacts of climate change in the two basins without adaptation. Based on these results, adaptation measures were identified for the two basins for different timeframes. Focus of the impact and adaptation assessment was on the base year and the year 2025. In the following subsection (Subsection 2.8.3) some results for the base year and for the year 2025 are presented.

2.8.3 Results

Introduction

The renewable water resources come from infiltration of rain and surface water. These quantities have been estimated on the basis of the level fluctuations measured by the piezometer[13] network. In addition, relationships between rain, infiltration and depth of water tables were established.

The renewable water resources in the aquifer of the zone amount to in the order of 8 billion m^3/year in periods with an average rainfall. The total water reserves include the resources of recently infiltrated rainwater combined with the resources formed in the last millennium and before.

Estimated runoff and available water resources

The estimated runoff for the two basins investigated is shown in Tables 2.7 and 2.8 for the baseline scenario and Tables 2.10 and 2.11 for the climate change scenario. Table 2.9 and Table 2.12 show the estimated water resources available with and without climate change.

Water needs

Table 2.13 shows that the accumulated water demand for the population, agriculture, livestock and other sectors is expected to increase from 1.20 billion m^3 in 2000 to 2.26 billion m^3 in 2025. The electricity sector has the highest demand and this is expected to increase from 5.65 billion m^3 in 2000 to 6.47 billion m^3 in 2025.

---

[13] A piezometer is a non-pumping well, generally of small diameter, for measuring the elevation of a water table.

Table 2.7. Results of estimations of river runoff (m³/s) by means of the RAINRU model for 2005–2025 in the Basin of Baoulé at Bougouni for the baseline scenario.

	Year				
	2005	2010	2015	2020	2025
January	0.90	0.90	0.90	0.90	0.90
February	0.90	0.90	0.90	0.90	0.90
March	0.90	0.90	0.90	0.04	0.90
April	0.90	0.90	0.90	0.90	0.90
May	0.90	0.90	0.90	0.90	0.90
June	15.22	17.44	19.59	21.74	23.89
July	87.80	93.68	99.38	105.07	110.77
August	182.69	192.79	202.57	212.34	222.12
September	230.59	242.74	254.50	266.25	278.01
October	184.87	194.02	202.88	211.73	220.59
November	100.27	105.46	110.48	115.50	120.52
December	26.54	28.04	29.49	30.93	32.38
Annual average	69.37	73.22	76.95	80.60	84.40

Table 2.8. Results of estimations of river runoff (m³/s) by means of the RAINRU model for 2005–2025 in the Sankarani at Sélingué for the baseline scenario.

	Year				
	2005	2010	2015	2020	2025
January	7.62	7.62	7.62	7.62	7.62
February	7.62	7.62	7.62	7.62	7.62
March	7.62	7.62	7.62	0.05	7.62
April	7.62	7.62	7.62	7.62	7.62
May	21.37	21.81	22.25	22.69	23.13
June	111.20	113.99	116.77	119.54	122.33
July	308.16	314.24	320.28	326.32	332.40
August	563.89	572.93	581.94	590.96	600.01
September	785.28	796.88	808.47	820.06	831.67
October	715.20	725.98	736.77	747.54	758.33
November	328.70	334.71	340.73	346.74	352.76
December	53.69	55.21	56.73	58.25	59.78
Annual average	243.17	247.19	251.20	254.59	259.24

Table 2.9. Total available water resources (10^9 m³) in the two basins for the period 2005–2025 in the baseline scenario.

	Year				
Water resource	2005	2010	2015	2020	2025
Renewable	12.65	12.76	12.97	12.97	13.08
Available	9.86	10.10	10.35	10.57	10.84
Total available	22.51	22.86	23.32	23.54	23.92
Reserves	82.1	82.1	82.1	82.1	82.1

Table 2.10. Results of estimations of river runoff (m³/s) by means of the RAINRU model for 2005–2025 in the Baoulé at Bougouni for the climate change scenario.

	Year				
	2005	2010	2015	2020	2025
January	0.98	0.98	0.98	0.98	0.98
February	0.98	0.98	0.98	0.98	0.98
March	0.98	0.98	0.98	0.05	0.98
April	0.98	0.98	0.98	0.98	0.98
May	0.98	0.98	0.98	0.98	0.98
June	4.18	3.27	2.55	1.27	0.98
July	62.69	60.49	58.24	55.84	49.38
August	147.11	144.67	142.12	139.50	128.92
September	189.86	188.46	185.52	183.76	171.40
October	156.85	156.56	154.78	154.21	145.31
November	84.38	84.82	83.89	83.96	80.13
December	21.49	21.74	21.37	21.46	20.56
Annual average	55.95	55.41	54.45	53.66	50.13

Table 2.11. Results of estimations of river runoff (m³/s) by means of the RAINRU model for 2005–2025 in the Sankarani at Sélingué for the climate change scenario.

	Year				
	2005	2010	2015	2020	2025
January	7.62	7.62	7.62	7.62	7.62
February	7.62	7.62	7.62	7.62	7.62
March	7.62	7.62	7.62	0.05	7.62
April	7.62	7.62	7.62	7.62	7.62
May	17.89	17.59	17.24	16.89	15.44
June	88.88	86.80	84.76	82.23	72.82
July	258.87	253.88	249.74	243.47	222.38
August	493.86	487.42	481.50	474.12	444.93
September	708.70	705.05	700.17	696.29	668.77
October	647.86	650.25	645.29	645.26	630.43
November	290.16	298.43	290.81	293.05	287.41
December	45.09	50.38	46.32	47.92	45.71
Annual average	215.15	215.02	212.19	210.18	201.53

Table 2.12. Total available water resources (10⁹ m³) in the two basins for 2005–2025 in the climate change scenario.

Water resource	Year				
	2005	2010	2015	2020	2025
Renewable	12.22	12.21	12.15	12.13	11.65
Available	8.55	8.53	8.41	8.32	7.94
Total available	20.77	20.74	20.56	20.45	19.59
Reserves	82.1	82.1	82.1	82.1	82.1

Table 2.13. Water demand (billion m³) in the two basins.

Sector	Year					
	2000	2005	2010	2015	2020	2025
Population	0.007	0.008	0.009	0.011	0.012	0.014
Livestock	0.011	0.013	0.016	0.019	0.022	0.027
Agriculture	1.18	1.33	1.51	1.73	1.94	2.22
Subtotal	1.198	1.351	1.535	1.76	1.974	2.261
Electricity production	5.65	6.18	5.30	5.59	5.30	6.47

So, the total water volume used for the socio-economic activities was 6.85 billion m³ in 2000 and will be 8.73 billion m³ in 2025, an increase of 27% in a period of 25 years.

It appears that renewable water resources can cover the needs of the main socio-economic activities up to 2025, as can be concluded from Fig. 2.11. However, the results of this global comparison can overlook the existence of local deficits during certain periods of the year. Hence, a more detailed analysis for the study areas needs to be carried out in the future. Estimated impacts without adaptation and adaptation measures in the different sectors are presented separately below.

The agricultural sector

A monthly analysis of crop water demand using the CRIWAR model (CNRST, 2003b) showed that, without adaptation, changes in monthly distribution of precipitation will have a negative impact on rainfed agricultural production. For maize this would lead to a decrease in production in the study areas, which is caused by a progressive reduction of the length of the rainy season. It would also entail a significant change in cereal production in the study areas. A reduction in the length of the growing cycle of these crops was found in certain areas, such as in Bougouni. This is because of the temperature rise as foreseen by the climate change scenario (CNRST, 2003a). The production and economic losses to cotton producers in the different study areas without adaptation are presented in Table 2.14,

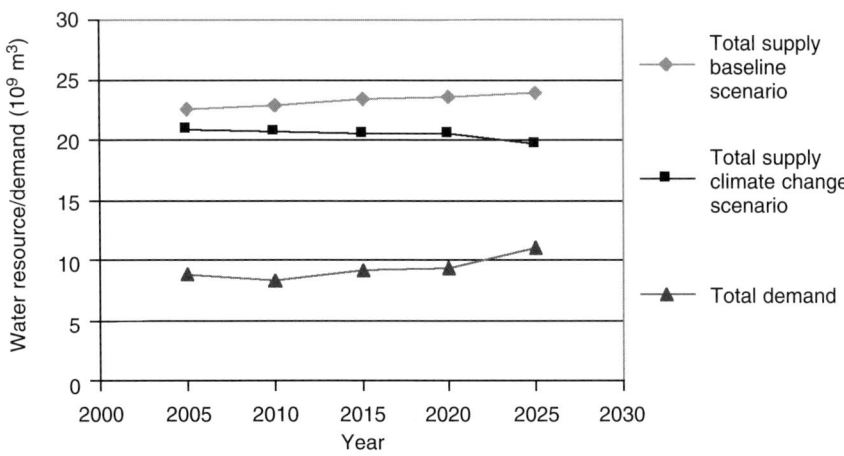

Fig. 2.11. Comparison of water supply for two scenarios (one without a climate change and one with climate change) and water demand by the socio-economic sectors in the basins of Sankarani and the Baoulé.

Table 2.14. Estimated cotton production and economic losses without adaptation in the study areas for 2005 and 2025 under the climate change and socio-economic scenarios.

Study area	Production losses (t) 2005	Production losses (t) 2025	Economic losses (1000 US$)[a] 2005	Economic losses (1000 US$)[a] 2025
Bougouni basin	150	450	51	153
Koutiala basin	200	1600	68	544
Dioila basin	1500	3500	510	1190
Yanfolila basin	370	750	126	255

[a] 1 US$ = 175 F CFA, 1 kg cotton valued at US$0.34 (2000 values).

demonstrating the impact of climate change on the production of cotton, an important export for Mali.

The most realistic adaptation measures must be focused on the control and management of water resources and the use of crop varieties that are less vulnerable to climate change.

The natural sources of water for rural communities (the main part of the population) and livestock are surface water and groundwater. As these water resources are expected to decrease with climate change (CNRST, 2003c), adaptation measures were identified to meet the demand for water. These include irrigation, water storage, water import, construction of wells and pools, raising awareness on good water management practices and improving livestock breeding practices. However, it is expected that financial resources of farmers are insufficient to pay for water transport and the realization of wells. Moreover, it is likely that people will migrate to other areas in times of water shortage as this is common practice for the people of Mali, especially for communities dependent on livestock. Under the climate change scenarios assessed water shortages could result in an exodus of people from the basins to urban centres and into neighbouring countries. Cattle will also be affected because the river that they were using for drinking water will dry up. The best solution for this would be the construction of a reservoir filled regularly with water.

Electricity production

Climate change could affect electric energy production by hydropower facilities because at least 70% of the national energy production is hydroelectricity. In order to minimize the effects of climate change, appropriate measures should be taken in time to secure the levels of electric energy production by hydropower facilities. Possible measures to assure electricity supply and demand include:

- the modification of the management plan for inflow and outflow of the Sélingué reservoir;
- the creation of more works to retain water allowing a better flow regulation of the Sankarani River upstream of the Sélingué Dam. This would avoid surpluses in water flowing away in the months September and October; and
- the reinforcement of the electricity production network by importing electricity, e.g. from the Manatali Dam (which concerns Mali, Senegal and Mauritania) or from other neighbouring countries such as Cote d'Ivoire.

Fisheries

Fisheries are bound directly to the abundance of water and therefore to rainfall. This is illustrated by the following example: the total fish capture in Mali is approximately 110,000 t during an average year of rainfall, as for example in 1966. In a dry year the captures drop dramatically, as for example in 1984 when only 54,000 t was caught, a decrease of nearly 50%.

The lake Sélingué Dam has a maximum storage capacity of 2 billion m^3 of water. This makes the restocking of capital in piscatorial products possible. A reduction in precipitation in this basin in the period 2005–2100, as predicted by the climate scenarios used in this study, will reduce the area of flooded surfaces and thus the amount of fish.

The reduction of fish quality and quantity will push fishermen to adopt other methods and tools for fishing. This might lead to an overexploitation of the resources and could lead to a change in the stock and a decrease in fish yield. As in the past during times of fish shortage, tensions can be expected to arise between communities because of the arrival of new fishermen by migration and because of the differences in the fishing methods of the different communities.

2.8.4 Experiences and lessons learned

During the studies several lessons were learned. Below some of the experiences and lessons learned are presented for specific areas.

Data availability and models

The analysis of climate change impacts in the domain of water resources requires the use of models. These models require data that are not always readily available. Even when these data sets exist they are often either insufficient or limited to local circumstances. As a result the study was performed with limited data sets for some areas. Because of the data limitations complex models could not be applied and therefore the results had to be based on simple models. It is important to note that the national experts that carried out the studies were not accustomed to some of the models, which had not been available before locally. Hence, assistance was needed from international experts. They were not always available, which led to delays in the implementation of the study. National experts were very committed to the project. For example, they participated in training sessions and spent more time on the project than they were obliged to.

Multidisciplinarity and interdepartmental project scope

Working in multidisciplinary and interdepartmental teams helped to overcome difficulties (such as human resource management difficulties, the needs of appropriate expertise) and promoted the exchange of knowledge and data between the different services (national technical services, private sector and NGOs) and the national team. Another merit of the multidisciplinary approach was that during the whole project, from developing scenarios until the analysis of the impacts of climate change on society, the experts in different fields worked together. This improved the insight on how climate change can influence water resources and how this will influence production (agriculture, livestock breeding, fishery etc.) and the population (food security, income, mobility, water bound conflicts).

2.8.5 Follow-up research

Results of this project form a good basis for future, more in-depth, assessments of climate change impacts on socio-economic systems in Mali. During different seminars several specialists expressed their interest in relation to the results of this project and considered initiating studies in other sectors, such as irrigated rice production and dairy production, including the quality of nutrition of milk and milk products. Research in the area of irrigated rice will need to be linked with research conducted in Niger, which has the potential to become one of the main irrigated rice suppliers in Africa. A study on the impacts of climate change on human health is now in progress together with experts from Burkina Faso.

2.8.6 Policy implications

The project contributed to raising political decision makers' awareness of climate change issues. Elements that grasped particular attention were the climate change impacts on cereal production and how these changes negatively affect producer incomes.

Also the general public, confronted with the study results, saw their relevance and the necessity of taking adaptation measures to ameliorate the impacts of climate change.

The political authorities appreciated the project results and wanted the different actors to take notice of the climate scenarios and the use of these scenarios for the development of the adaptation strategies. First actions in this regard were taken during the exploratory consultations for the identification of activities for the NCCSAP Phase 2. All actors present were unanimously positive on the quality of the results of the first phase. They agreed on the necessity to capitalize results of this phase in the second phase. One of the first people to appreciate the importance of this work was the Chief of Staff (Vice Minister) of the Ministry of the Environment. This illustrates the interest of Malian authorities in climate change impacts and adaptation.

2.8.7 Conclusions

The use of simulation models requires a compilation of different data. Moreover, the quality of the model results depends on the quality of the data input. The lack of data for some locations influenced the choice of the study areas and the models that could be used. Hence, results should be seen as first order estimates.

Climate change impacts on water resources and agricultural production in the basins of Bougouni, Dioila, Yanfolila, Sélingué have been determined for different timeframes (2005 to 2025). Additionally, a first order assessment of climate change impacts on hydroelectricity production in the Sankarani basin, was made.

It is concluded that climate change will lead to a decrease in rainfall and an increase in temperature in Mali. The decreases in rainfall will lead to significant water shortages in specific areas. In areas with severe water shortages the local population and policymakers will have to take adaptation measures to cope with climate change. Local measures could include importing water, water and crop storage for the delayed use of water resources and stocks of foods, improved water resource management and digging wells. Costs of these activities can be expected to be a limiting factor. Hence, climate change can be expected to lead to large groups of people migrating to areas where water would be available, to the urban centres and to neighbouring countries if governments and other stakeholders do not take appropriate adaptation measures. Adaptation measures identified in this study include:

- the realization of small-scale dams, construction of reservoirs to retain water and the creation of other new systems to retain surface water;
- measures to secure the income of the population. This will involve changes in agriculture, fisheries and market cultures. Policies to restrain the use of water resources and mobilize the people of Mali; and
- policies to mobilize other partners, such as NGOs, the private sector and bilateral and multilateral partners etc.

More in-depth studies are needed to identify the most appropriate adaptation measures and strategies. The results of this project provide a good basis for these future studies. The government of Mali is very interested in the issue of climate change and acknowledges the need for adaptation measures to be taken. Indeed, for a country such as Mali, situated in arid and semi-arid zones in Africa, adaptation to climatic change is a question of survival and not of choice.

Main Publications From the Research

CNRST (2003a) *Elaboration of Climate Change Scenarios for Mali*. Project Climate/Study 1. National Centre for Scientific and Technological Research (CNRST), Bamako, Mali, 96 pp.

CNRST (2003b) *Vulnerability and Adaptation of the Maize and the Cotton at the Climate Change in Mali*. Project Climate/Study 2. National Centre for Scientific and Technological Research (CNRST), Bamako, Mali, 106 pp.

CNRST (2003c) *Vulnerability and Adaptation of Water Resources at the Climate Change in the Basins of the Sankarani and of Baoulé*. Project Climate/Study 3. National Centre for Scientific and Technological Research (CNRST), Bamako, Mali, 126 pp.

2.9 Mongolia

Punsalmaa Batima[14]

2.9.1 Introduction

Climate change and extreme climatic events could have dramatic impacts on Mongolia's economy and natural systems, with the potential in some cases for irreversible damage to ecosystems. Agriculture, including crop and livestock production, water and forest resources, as well as biodiversity, are among the most vulnerable systems.

The Mongolian government perceives that any negative effects that may be experienced from climate change could be equivalent to a national disaster. Extreme climatic events, with or without a superimposed change in climate, could have tremendous adverse effects on the Mongolian economy and on the well-being of its people.

Moreover, Article 4 of the United Nations Framework Convention on Climate Change (UNFCCC) calls upon all Parties to prepare a National Action Plan on climate change that describes greenhouse gas (GHG) inventories and actions to reduce GHG emissions and to adapt to climate change. In addition, a National Action Plan is necessary not only to meet the country's obligations under the UNFCCC, but also to set priorities for action and integrate climate change concerns into other national environment and development plans. Therefore, Mongolia requested assistance from the government of the Netherlands in the preparation of its National Action Plan on Climate Change (NAPCC).

The Mongolia study under the NCCSAP aimed to cover:

1. Analysis and evaluation of GHG mitigation options.
2. Impact and adaptation assessment of the agricultural sector (crop production).
3. Impact and adaptation assessment of rangeland and livestock sectors.
4. The preparation of the NAPCC.

Hence the project expert team has completed impact and adaptation assessments for the natural resources that depend on water resources: permafrost, soil, grassland and agricultural activity such as livestock and crop production, as well as mitigation analysis of GHGs. The NAPCC was prepared on the basis of the results of these studies. Water is a cross-cutting issue which is important for crop production, livestock husbandry and domestic use.

Water is crucial for the maintenance and restoration of the environmental carrying capacity of pasture/grassland and other economic activities such as crop growth. Thus water must be an integral part of the total environmental management package. We believe that through water resources management, land use and human activities can be guided and controlled in Mongolia. Therefore, in this section we will focus on water.

Nearly half of the population lives in rural areas where people use wells, rivers, springs and other surface waters for daily use. The provincial centres also have a strong rural character, i.e. most households grow crops and rear animals. Therefore, both urban households and herdsmen depend on the availability of water. But freshwater is a vulnerable resource. In some areas of the steppe and the Gobi, groundwater is the only source for daily use. Different types of wells (artesian, pipe, tube or hand) serve as a water source for residential use and for livestock watering. Rivers are the main source of water for residents of mountainous regions. Herdsmen must decide to rest at places where water is available or where the herds can graze. For instance in the Gobi, herdsmen's families move six or seven times a year to find a place where both water and pasture are available. The movement around water resources leads to increased pressure on the land in the vicinity of these scarce resources. The resulting high cattle densities in turn lead to overgrazing, trampling, erosion and sand movement. Accordingly water shortage has become one of the major socio-economic problems, especially in the Gobi and the steppe.

---

[14] Institute of Meteorology and Hydrology, Hudaldaany gudamj-5,Ulaanbaatar-46, Mongolia. Tel.: 976 11 318750, Email: mcco@magicnet.mn

2.9.2 Approach

Climate change scenarios

The scenarios of the General Circulation Models (GCMs) were used to examine the impact of increased GHG concentrations on the future climate of Mongolia. The results of GCM scenarios have been taken from the database of the Intergovernmental Panel on Climate Change (IPCC) Working Group I. We used the scenario of IS92a, which considers only GHG concentrations. The five scenarios selected for current study are: (i) the Max Planck Institute of Hydrology and Meteorology of Germany (ECHAM4) model; (ii) the Canadian Climate Centre Model (CCCM); (iii) the UK Meteorological Office and Hadley Centre (HADLEY) model; (iv) the Geophysical Fluid Dynamics Laboratory (GFDL) model; and (v) the Division of Private Atmospheric Research, Australia (CSIROKM2) model. The results of GCM scenarios were given in grid boxes of $2.8° \times 7.5°$. Therefore, in order to find the changes in temperature and precipitation at the study location, use is made of two-dimensional interpolation. The values of temperature and precipitation are determined on grid boxes of $0.5° \times 0.5°$.

Sensitivity analysis

Here hypothetical and systematic changes in temperature and precipitation are used to test the sensitivity of the outcomes with respect to these two inputs.

Water resources study

The Basin Conceptual Model (BCM) was used to assess river runoff under climate change. The BCM is a balance model that uses a monthly time step and requires multi-annual monthly mean values of precipitation, temperature, potential evapotranspiration and runoff as input. It takes the storage of the previous month to compute infiltration, evapotranspiration and runoff. It contains six parameters, with two of the parameters being the upper and lower temperature bounds on the freezing and melting process (Batima, 1995; Batima *et al.*, 1996).

A split sample test was used to evaluate the hydrologic model. In this test the historic record is broken into two segments, one used for calibration and the other for validation. Since the ranges of climatological and hydrological data for the selected rivers were different, the last 24 years are selected for model verification and testing. The first 10 years (1972–1981) were used for calibration and the remaining 14 years (1982–1996) were used for validation. The estimation of river runoff is done for each hydrological year. The hydrologic year begins in October, when snow accumulation is assumed to be zero. The estimated condition of soil moisture and snow pack at the end of each year is then taken as the initial condition of the calculation for the next year. The correlation coefficient and the average monthly error are used to describe model performance.

2.9.3 Results of the water resources assessment

Introduction

It is estimated that water resources in Mongolia amount to 34.6 km^3 of which 83.7%, 10.5% and 5.8% account for lakes, glaciers and rivers, respectively (Myagmarjav and Davaa, 1999). The total annual water use is 0.5 km^3 (Khuldorj, 1999). Hence, water use seems small compared to the water resources. However, water resources are unequally distributed over the country, i.e. in the northern part of the country the available water per capita is four to five times more than the world average while it is ten times less in the southern part, i.e. in the Gobi (Myagmarjav and Davaa, 1999).

Twenty river basins of different scale and climatological characteristics were selected. Selection criteria included basin size, varying climatic and basin characteristics (hydroclimatic zones), as well as time series data availability. The climate baseline of water resources was based on the monthly mean temperature, the monthly precipitation, and the duration of sunshine hours for the selected 20 rivers.

Sensitivity analysis

With the set of hypothetical scenarios we have generated a series of results that gave insight into the sensitivity of the basin flows to climate variations.

As expected, river runoff is much more sensitive to precipitation changes than to temperature changes. The analysis indicates that if annual precipitation drops by 10%, while the temperature remains constant, the average river flow is reduced by 7.5%, 12.4% and 20.3% in the Internal Drainage Basin, the Arctic Ocean Basin and the Pacific Ocean Basin, respectively. If, besides the precipitation drop, average temperature increases of 1, 2 and 3°C are taken into account, an additional flow reduction of 3–11% is expected. In other words, it appears that for each °C temperature increase, there is a 2–6% annual flow decrease.

Table 2.15 shows the results of the sensitivity analysis for the three basins mentioned above. As can be seen from the table, the Pacific Ocean Basin is more vulnerable to changes in precipitation and temperature than the other two basins. Moreover, the sensitivity analysis shows that the river flow in the Internal Drainage Basin rivers *increases* with increasing temperatures under constant precipitation levels. Since a glacier is the major source of river runoff for most rivers in the Mongol Altai and high altitudes in the Hangain Mountains, and because the increased temperature would result in intensification of glacier melting, river runoff is likely to increase even when precipitation remains the same.

The impact assessment

With similar scenario assumptions and driving forces, climate projections can vary considerably among GCMs. Therefore, to allow for critical comparisons, we analysed a range of climate projections by five different GCM outputs. The results reflect the high level of uncertainties in the results of different climate models. Nevertheless, despite their limitations these model results provide a first indication of the general range of possible flow alterations due to climate change.

CCCM. According to this scenario the runoff of the rivers will increase by 18–124% in 2040. The highest increase is expected to occur in the rivers flowing from the southern slope of the Hangain Mountains and the lowest in the rivers in the Uvs Lake Basin. However, the runoff of the Muren River that flows from the Huvsgul Mountain will decrease by 0.3%. In 2040, the runoff of the Internal Drainage Basin, the Arctic Ocean Basin and the Pacific Ocean Basin, seems to increase by 56%, 50%

Table 2.15. Changes (%) in river runoff at the three basins investigated.

Changes in temperature	Changes in precipitation				
	−20%	−10%	0%	+10%	+20%
Internal Drainage Basin					
0°C	−20.2	−7.5	0.0	21.2	37.3
+1°C	−17.3	−5.9	6.1	26.2	40.0
+2°C	−20.2	−8.9	3.7	21.3	36.3
+3°C	−22.8	−12.1	0.2	17.0	30.6
Arctic Ocean Basin					
0°C	−22.3	−12.5	0.0	12.8	27.4
+1°C	−23.5	−14.1	−3.8	9.1	22.9
+2°C	−27.2	−18.4	−9.3	1.9	14.5
+3°C	−29.0	−21.6	−13.5	−3.0	7.9
Pacific Ocean Basin					
0°C	−29.3	−20.3	0.0	17.7	31.2
+1°C	−36.9	−26.9	−15.1	5.0	21.2
+2°C	−40.1	−31.9	−20.7	−5.5	8.2
+3°C	−42.1	−31.6	−23.1	−9.8	6.1

and 65%, respectively. The CCCM result indicates substantial *decreases* in the water resources during the period 2040–2070.

GFDL. In this scenario the runoff of the rivers will increase by 10–90% in 2040. The highest increase seems to occur in the Pacific Ocean Basin and the lowest in the Arctic Ocean Basin. The river runoff in the period of 2040–2070 is expected to be 2–24% higher compared to the current runoff, and 15–38% lower compared to 2040. In sum, the river discharges in Mongolia are expected to increase until 2040 and most likely to decrease in the period 2040–2070.

ECHAM4. In this scenario, the river runoff is expected to change from *minus* 25.1% to plus 114% in 2040. The increase in river flow would occur in all rivers in the Arctic and Pacific Ocean Basins. There is going to be a decrease in river flow of 10–25% in the rivers flowing from the northern slope of the Mongol-Altain Mountain and the rivers in the Uvs Lake Basin. This scenario indicates that the water resources in the Arctic Ocean Basin would remain almost the same in the period of 2040–2070 compared to 2040, while in the other two basins the water resources seem to decrease by about 18%.

HADLEY. According to this scenario the river runoff is likely to increase by 3–65%. The highest increases seem to occur in the Pacific Ocean Basin and the lowest in the Arctic Ocean Basin, which was also found in the other scenarios. The river flow appears to continue to increase in the period 2040–2070 in the Arctic Ocean and the Internal Drainage Basins by about 11% compared to 2040, whereas the river flow in the Pacific Ocean Basin is expected to decrease by 19%.

The scenarios show that Mongolia's water resources are most likely to increase until 2040 and then decrease until 2070 towards the current levels. For example, in the Internal Drainage Basin the river runoff is expected to be 5–120% higher in 2040 compared to the current runoff. Especially CCCM and CSIROKM2 give very high increases in river runoff. But after 2040, runoff starts to decrease again. The river flow patterns strongly follow the temperature and precipitation patterns. According to the climate change scenarios after 2040, temperature is expected to continue to increase, while precipitation starts to decrease. If this pattern continues, it is clear the water resources will continue to decrease from 2040.

The scenario results also show that global warming will not significantly influence the seasonal distribution of river flows (Fig. 2.12). About 60–80% of the annual water runoff is formed during the rainy season. About 80% of all rainfall comes down in two to four intensive showers. They cause high flows, which run off quickly and their contribution to water storage is therefore negligible. From the sensitivity analysis we found that the river flows are more dependent on precipitation than on temperature.

General adaptation measures

Taking into account the existing scarcity of natural water supplies and their anticipated alterations, some adaptation measures we should consider are:

- The results from the climate change analysis as presented above show future flow alterations. Whatever scenario is considered, the presented climate change analyses strongly suggest that significant effects in the form of altered flow regimes, regardless of decreasing or increasing trends, may put additional pressure on the river water system. In consequence, a prospective strategy for water resources management needs to include the entire range of climate change scenarios. Hence, it is clear that traditional water resources management and development approaches and technologies can no longer provide sustainable water resources. Accordingly, new, integrated approaches are needed to reconcile conflicting interests on the use and the conservation of water resources. Thus, one of the most important adaptation measures is to prepare our next generation of water managers and to create the scientific, technical, institutional and organizational capacity to deal with the effects of climate change. To elaborate a comprehensive management strategy and to train national experts, it is necessary to implement national and international

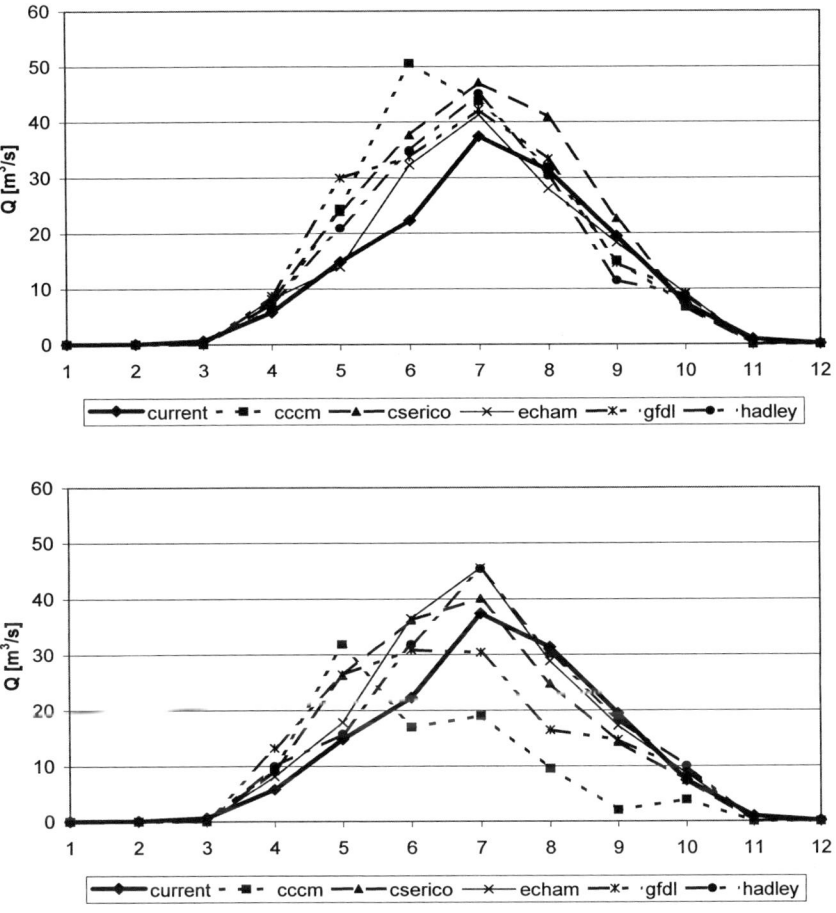

Fig. 2.12. Current and simulated monthly runoff (Q) by the five GCM scenarios at the Tuul River in 2040 (top) and 2070.

projects funded by the government and international organizations.
- It is important to apply a water resources management policy that focuses on the rational use of available water resources without exceeding their renewal rate. In particular, the conservation of freshwater ecosystems in runoff formation zones is an essential element for water resources management, because these areas largely determine the sustainable recharge of groundwater and surface water storage. In order to reach this goal, it is necessary to incorporate the upper areas of the basin runoff where most of the water originates.
- Water quality may be more important than quantity since available water often does not meet drinking water standards. Recently, most of the river waters were assessed as fresh by the water quality requirements (Batima, 1998). There are several important options, such as the improvement of the operation of wastewater treatment plants to reduce pollution, maintenance of sanitation

zones to protect river water quality, development of water recycling technology, organization of chemical as well as biological monitoring to control and predict water quality, etc.
- The Dublin International Conference on Water and Environment in 1992 (organized by the World Meteorological Organization) emphasized that water should be considered as an economic good (see also Subsection 4.3.2). Water use/supply for animal watering and irrigation is still free in Mongolia. People do not care about commodities that are supplied for free. Therefore, the issues of conservation of water resources and demand side management should be highlighted in the future through public awareness and training. The intensity of precipitation is going to increase due to climate change and, consequently, flood events will increase. At the same time, in low precipitation areas, the drought occurrence might increase. In this respect, studies on the physical condition and prediction of such events can be a significant adaptation option to overcome the possible hazardous impacts of climate change. Waters in the Gobi desert are naturally saline and will become more saline under increased temperature. Hence, water softening and purifying equipment is needed to provide a safe water supply.

2.9.4 Experiences and lessons learned

Working with models and scenarios

The Water Balance Equation was found to be an efficient tool for the climate-water resources impact assessment (Zdzislaw, 1996). After consideration of a number of standard models such as CLIRUN3, WATBAL and the BCM (Yates and Strzepek, 1994), the latter was selected for the study. One distinguishing characteristic of Mongolian rivers is that they are all covered by thick ice layers of about 1.0–1.8 m for 5 or 6 months a year and small rivers are even frozen to the bottom. The initial form of the BCM did not give good results. Therefore, we have made some modifications (adding coefficients for snow melting and surface runoff), which resulted in better estimations of snow accumulation and basin runoff for both seasons (Batima, 1995; Batima et al., 1996). Hydrological, meteorological and socio-economic data were used for water resources assessment. Although the length of the time series was rather short to show the trends of water resources of past climate change, it was sufficient to model the river runoff.

The simulation results obtained with climate scenarios were compared with the actual observation data. It was found that the estimated runoff exceeds the observed one by 0.01–5.5% in most rivers. In a few rivers, like Sagsai, Ylz, Herlen and Yrdtamir, it is underestimated by 0.1–3.4%. In spite of these errors, the estimated runoff has a good correlation with the observed one for each selected river basin. The correlation coefficient between simulated and observed runoff was between 0.66–0.86 and the average monthly error between 0.27–7.49. These statistical values are good enough to conclude that the BCM performs well in exploring the current situation and that it can be used for simulation of river runoff with climatic variability. The high significance levels of the statistical values derived during calibration and validation can be explained by assumptions made on the basis of the nature of the monthly mean runoff, i.e. the strong seasonality with high flows during the rainy season and the freezing of the rivers in the winter.

Multidisciplinary research in the studies

The Technical Expert Team (TET) included experts from relevant sectors including government agencies, academic institutions and private companies, including the Ministry of Nature and the Environment (MNE), the National Agency for Meteorology, Hydrology and Environment Monitoring (NAMHEM), the Institute of Meteorology and Hydrology (IMH), the Ministry of Agriculture and Industry (MAI), the Ministry of Health and Welfare (MHW), the Ministry of Infrastructure and Development (MID), the Institute of Geo-Ecology (IGE), the Institute of Biology (IB), the Mongolian National University (MNU) and the University of Agriculture (UA).

The TET divided itself into sub-teams,

namely, the team for preparation of the NAPCC, the team on analysis and evaluations of the GHG mitigation options, the team on impact and adaptation assessment in the agriculture sector (crop planting) and the team on impact and adaptation assessment in the rangeland and livestock sectors. However, we recognize that the constitution of TET was biased towards the scientific community and that we should also have invited some NGOs to get additional inputs.

Limitations

The spatial grid of the GCM outputs is relatively coarse compared to the river basin scales. Therefore, it could be that the modelled outcomes are relatively uncertain. Furthermore, a prospective strategy for water resources management needs to include the entire range of climate change scenarios.

The BCM was used to estimate changes in water, it does not provide detailed analysis of micro-level processes, which may be where most of the impacts of climate change occur. The model does not provide information on floods, droughts or other extreme events. Also, groundwater impacts are not addressed in regard to impacts on recharge or as an alternative to surface resources. The melting of glaciers was also not taken into account with this method.

Adaptation options have been evaluated mainly on the basis of expert judgement. Therefore, we are aware of the small scientific basis of the discussion and as a result the indicated adaptation options were rather subjective. A screening matrix to examine the adaptation options and evaluate their suitability for implementation has been used for the first time in the study (Batima and Dagvadorj, 2000).

The study created a framework for subsequent mainstreaming of climate change issues in the national policy and legal framework. It has enhanced capacities in the scientific and research community of Mongolia to appreciate climate change issues and further work on them in the context of its results. The project has further highlighted the need for a stronger effort of creating awareness among the stakeholders and decision makers.

2.9.5 Follow-up research

The results presented here cover the changes of river runoff. The other hydrological parameters such as river ice trends and river flow changes in different water regimes have not been studied. Thus the results should be considered as a starting point for more detailed discussions about anticipated climate and global change effects on not only water resources but also freshwater ecosystems. Additional studies are required to validate the results. From an ecological perspective, it is important to identify quantitative thresholds that mark the limits of alteration to which an ecosystem can adapt. In this context the Mongolia project has been able to create a solid foundation for further work on scientific and policy issues. It clearly defined the issues that are relevant within the national context, and identified potential areas for further research.

After the completion of the NCCSAP project, Mongolia further developed the project under initiation of extended financing of the Global Environment Facility (GEF) Climate Change Enabling Activity (Part II). This facilitated Mongolia to build additional capacity in priority areas in order to contribute to the implementation of the NAPCC. The project focus was an assessment of technology transfer needs in the Mongolian energy sector, public awareness and education, and enhancement of national capacities to prepare national communications.

Through implementing this project Mongolia has achieved:

1. An assessment of technology transfer needs in the Mongolian energy sector. Besides training engineers and managers in industry and heat only boilers (HOBs), a technology needs assessment was done in more than 100 industries in food processing, wool and cashmere manufacturing, building material production and textile production as well as in other sectors. As a result of this work, which concerned inefficient HOBs, the government included this as one of the pillars for its Strategy to Donors at the Consultative Group meeting in July 2002. As a follow up, the International Donors Agency is identifying components for financial support.

2. Public awareness and education. Three books, *Climate Change, Greenhouse Gas and its Effects* and *Climate Change and Sustainable Development* have been written and published in Mongolian for undergraduate, graduate and postgraduate students, and two information kits, *GHG Emission and its Mitigation* and *Climate Change*, were published in Mongolian for a wider public audience.

3. The enhancement of national capacities to prepare national communications. A national climate change website was established. It was designed for the distribution of the information and results that were produced in past and ongoing climate-related activities in Mongolia. The most important books, brochures, information kits, posters, articles and presentations that were produced have been published on this website.

Research on 'Potential Impacts of Climate Change and Vulnerability and Adaptation Assessment for Rangeland and Livestock in Mongolia' is currently being carried out within the Assessments of Impacts of and Adaptations to Climate Change in multiple regions and sectors (AIACC) programme.

2.9.6 Policy implications

The government of Mongolia approved the NAPCC on 19 July 2001. The Ministry for Environment and Nature organized a National Workshop on Climate Change Issues in Mongolia on 27–28 June 2002 in Ulaanbaatar. The objective of the workshop was to draw the attention of stakeholders at different levels (policy and decision making, research and educational institutions, NGOs, private sector and the general public) to the implementation of NAPCC and disseminate and discuss climate change issues such as GHG emissions and mitigation, expected climate change impacts, and vulnerability and adaptation in different sectors of the country. One hundred and sixty delegates participated including eight members of parliament, 33 officers from different ministries and agencies, 42 delegates from 21 provinces, 26 delegates from research and educational institutions, 40 delegates from private companies and NGOs and 11 journalists. Box 2.4 presents a citation of a part of the Minister's speech at this workshop.

As a follow-up to this national workshop, local and sectoral workshops on climate change issues have been organized. In particular, a regional workshop in the eastern province was organized by local governors and local specialists. The main goal of this workshop was to discuss how to integrate climate change issues in regional development.

Similarly, a workshop titled 'Climate Change and Agriculture' was organized by organizations such as the Plant Science and Agriculture Training and Research Institute at Darkhan, the Institute of Meteorology and Hydrology, the Association for Agriculture Farmers and Flour Producers, the Canadian Agroteam Co. Ltd, the Agriculture Development Fund and the National Centre for Technology Renovation in Agriculture. Participants discussed how to set the adaptation options included in the NAPCC into arable land development.

Box 2.4. Citation from the Minister at the National Workshop on Climate Change Issues.

Almost one third of Mongolia is defined as a very vulnerable region under climate change. All rivers in this region are seasonal, i.e. they flow only during the rainy season. Nearly 90% of the lakes have an area of less than 1 km^2 and strongly depend on precipitation. The residential water demand can be met by groundwater in towns, villages and settlements, but the pasture water supply will be the most difficult problem to solve, particularly in arid and semi-arid areas. Therefore, further study of water resources under climate change should focus on this vulnerable area.

2.9.7 Conclusions

There are four main climatic water regimes observed in the rivers of Mongolia. These are: winter low flow period; spring runoff period due to snow melting; summer runoff period due to rainfall; warm season low water period. Due to the unequal distribution of precipitation and the sharp changes of seasons, the water resource varies significantly across time and space.

The sensitivity analysis indicates that if annual precipitation drops by 10%, while temperature remains constant, the average river flow reduces by 7.5%, 12.4% and 20.3% in the Internal Drainage Basin, the Arctic Ocean Basin and the Pacific Ocean Basin, respectively. If, besides the precipitation drop, average temperature increases of 1, 2 and 3°C are taken into account, an additional flow reduction of 3–11% is expected. In other words, it appears that for each °C temperature increase, there is a 2–6% annual flow decrease.

The results of the GCM-based scenarios show that Mongolia's water resources are most likely to increase until 2040 and then decrease until 2070 towards the current levels. For example, in the Internal Drainage Basin the river runoff is expected to be 5–120% higher in 2040 compared to the current runoff. CCCM and CSIROKM2 especially give very high increases in river runoff, but after 2040, runoff starts to decrease again. The river flow patterns follow the temperature and precipitation patterns. According to the climate change scenarios after 2040, temperature is expected to continue to increase, while precipitation starts to decrease. If this pattern continues, it is clear the water resources will continue to decrease from 2040. This suggests that it will be worthwhile taking water storage in the period 2000–2040 into account.

The GCM scenario results are confirmed by the sensitivity analysis results: especially at high altitudes, the river flows tend to increase with temperature if precipitation is unchanged. This is most likely expected to happen during the first decades of the global warming scenarios.

Surface water resources are expected to decrease after 2040. The rate of decrease is higher in the eastern part than in the western part. This may be caused by changes in natural zones and permafrost shift. The cool temperature zone in the Hentein mountain region will be reduced drastically in 2040 and will be replaced almost completely by the warm temperature zone in 2070. The permafrost boundary will rapidly shift to the north and permafrost will nearly disappear by 2070.

Projected increases in air temperature will lead to severe changes in snow and glacier melt. The effects of these changes are highly difficult to predict. It also should be kept in mind, however, that the melting of the glaciers cannot be considered a sustainable process, but rather represents a non-renewable water resource that is limited by the individual glacier volumes. The effect could be increased average flows, increasing until the entire glacier is melted, then decreased flows. This may cause earlier and intensified high flows from spring snowmelt followed by extended and more pronounced low flow periods.

Main Publications From the Research

Batima, P. (1995) Impact of climate change to the water resources of Mongolia. In: *Climate Change in Arid and Semi-arid Regions of Central Asia*. Proceedings of the second Mongolian–China Symposium, September. Ulaanbaatar, pp.114–124.

Batima, P. (1998) Water quality assessment in river water chemistry and water quality assessment in Mongolia. PhD thesis, Institute of Meteorology and Hydrology, Ulaanbaatar, Mongolia, 128 pp.

Batima, P. and Dagvadorj, D. (2000) *Climate Change and its Impacts in Mongolia*. JEMR Publishing, Ulaanbaatar, Mongolia, 227 pp.

Batima, P., Bayasgalan, S., Bolortsetseg, B., Dagvadorj, D., Migiddorj, R. and Nathagdorj, L. (1996) Vulnerability and Adaptation Assessment for Mongolia. Vulnerability and Adaptation to Climate Change. In: Smith, J.B. *et al.* (eds) *Interim Results from US Country Studies Program*. Environmental Science and Technology Library, Ulaanbaatar, Mongolia, pp. 183–207.

Batima, P., Dagvadorj, D. and Dorjpurev, J. (2000) *Green House Gas Mitigation*

Potentials in Mongolia. JEMR Publishing, Ulaanbaatar, Mongolia, 227 pp.

Batjargal, Z., Dagvadorj, D. and Batima, P. (2000) *Mongolia National Action Programme on Climate Change.* JEMR Publishing, Ulaanbaatar, Mongolia, 156 pp. (in English and Mongolian).

Institute of Meteorology and Hydrology (2001) *Initial National Communication of Mongolia.* Institute of Meteorology and Hydrology, Ulaanbaatar, Mongolia, 110 pp. Available from: http://unfccc.int

Khuldorj, B. (1999) *Mongolian Action Programme for the 21st Century.* ADMON Publisher, Ulaanbaatar, Mongolia.

Myagmarjav, B. and Davaa, G. (1999) *Surface Water of Mongolia.* Interpress, Ulaanbaatar, Mongolia, 267 pp.

Yates D. and Strzepek K. (1994) *Comparison of Models for Climate Change Assessment of River Basin Runoff.* WP-94-45, IIASA, Laxenburg, Austria, p. 25.

Zdzislaw, K. and Krasuski, D. (1991) *Sensitivity of Water Balance to Climate Change and Variability.* WP-91-047, IIASA, Laxenburg, Austria, 25 pp.

2.10 Senegal

I. Niang-Dop, M. Dnsokho, A.T. Diaw, S. Faye, A. Guisse, I. Ly, F. Matty, A. Sene,[15] P.S. Diouf,[16] K. Gueye[17] and P. Ndiaye[18]

2.10.1 Introduction

Senegal was among the first non-Annex I Parties of the United Nations Framework Convention on Climate Change (UNFCCC) to present its Initial National Communication in 1997 (République du Sénégal, 1997). The first study, which started in 1990, assessed the impacts of sea level rise (SLR) on the Senegalese coastline (Dennis *et al.*, 1995). The second study was part of country studies on global food security and climate change conducted in Zimbabwe, Kenya, Senegal and Chile (Downing, 1992). These two studies were funded by the US Environmental Protection Agency. The choice of Senegal was in fact determined by the results of a first Global Vulnerability Analysis that was made by the Intergovernmental Panel on Climate Change (IPCC) Coastal Zone Management Subgroup to test the Common Methodology relative to the impacts of climate change in the coastal zones (IPCC/CZMS, 1991a). In this study Senegal was ranked as the eighth most vulnerable country to SLR among 181 countries (Hoozemans *et al.*, 1993). Senegal was thus chosen to conduct pilot studies on the impacts of climate change on its coastal zone and on agriculture.

In Senegal, three studies were conducted under the NCCSAP: the development of a national climate change scenario (Gaye *et al.*, 1998) and two vulnerability and adaptation studies for agriculture and coastal zones. This choice for the sectors was determined by their economic importance for the country as well as by the existing expertise in the country. Here the results of the vulnerability and adaptation study for the coastal zone are presented because it is a relatively comprehensive study.

During the same period three other vulnerability and adaptation studies were developed: one on water resources, one on fisheries and the last one on tourism as part of the United Nations Institute for Training and Research (UNITAR) Climate Change Training Programme (CC:TRAIN). It is important to note that Senegal developed an implementation strategy for the Climate Change Convention in 1999 (République du Sénégal, 1999).

The 1990 coastal zone vulnerability and adaptation study only assessed the impacts of different scenarios of SLR – namely coastal erosion and inundation – on the coastline with rough estimations of population and land value at risk as well as an evaluation of the costs of

---

[15] University Cheikh Anta Diop of Dakar. Department of Geology, Faculty of Science, Dar-Fann, Senegal. Tel.: +221-8250736, Fax: +221-8246318, Email: isabelle@enda.sn
[16] World Wildlife Fund, Senegal.
[17] Centre Expérimentale de Recherche et d'Etude pour l'Équipement (CEREEQ).
[18] Ministry of Environment, Senegal.

adaptation, mainly protection. Within this context, the main purpose of the NCCSAP-funded study on the impacts of climate change on the Senegalese coastline was to develop a more comprehensive and integrated picture of the coastal vulnerability to climate change.

2.10.2 Approach

Based on the results and experiences of the 1990 vulnerability and adaptation study, it was decided to include the following improvements in the NCCSAP-funded study:

- Consider case studies that are representative of the main coastal environments. Two geographic locations were chosen: the Cap Vert peninsula, which is an example of an highly urbanized coastal zone and of sandy coasts; and the Saloum estuary, which is representative of a mangrove estuary – the second major coastal environment in Senegal;
- Use a multidisciplinary approach by building a team constituted of 11 Senegalese experts from different fields;
- For each case study, consider not only the impacts of SLR but also other potential climate change impacts such as changes in rainfall, temperature and upwellings;
- Perform a more comprehensive assessment of socio-economic impacts by using socio-economic scenarios and discount rates;
- Have a broader approach of adaptation to enlarge the basic solutions developed by the IPCC/CZMS: retreat, accommodate, protect (IPCC/CZMS, 1991a).

The study used a mix of methodologies ranging from simple expert judgement to the use of simple and complex models, such as the Bruun rule to calculate the land losses due to coastal erosion and the FEFLOW model to evaluate the impacts of climate change on the position of the saltwater intrusion in coastal aquifers (Faye et al., 2001). The IPCC methodology (Carter et al., 1994) and the IVM/UNEP *Handbook on Methods for Climate Change Impact Assessment and Adaptation Strategies* (Feenstra et al., 1998) were used as basic methodologies in the project.

2.10.3 Results of the coastal zone study

In this section we will first give an introduction to the two case study areas. Then the main results of the vulnerability and adaptation study will be presented.

The Cap Vert peninsula

The geographical delimitation of this case study was made to include three components: (i) the metropolitan area of Dakar; (ii) part of the north coast, characterized by major dune systems inside which are interdune lows, the *niayes*, partly inundated by the coastal aquifer; and (iii) part of the south coast where the sandy coast is bordered by a single barrier system and occupied by a number of villages and small towns. The total area considered is 1597 km^2. In 1992, the population was 2.3 million inhabitants representing 30.5% of the total population and 60% of the coastal population. During this year, the activities in the study area contributed 37% of the national GDP, mainly from services and industrial sectors. Economic activities linked to the coastal zone are fisheries and market gardening that takes place in the *niayes*.

The Saloum estuary

This is a complex estuarine system comprising three main rivers: Saloum, Diomboss and Bandiala with numerous small tidal channels. All of them are bordered by mangrove forests along their downstream parts. The Saloum River is bordered by a long sand spit: the Sangomar sand spit is 15 km long. The total area covered by this second zone is 4309 km^2, most of it 2 m below sea level. In 1992, the total population in the estuary was 540,000 inhabitants (7.2% of the national population and 14% of the total coastal population). This population is distributed in two middle-sized towns (Kaolack and Fatick) and a great number of small villages. The main activities in the rural areas are fisheries, agriculture and tourism. It was estimated that these activities in the estuary contributed 12.3% of the national GDP in 1992.

Main biophysical impacts

In the two case studies, it was possible to determine land losses due to coastal erosion and inundation but also impacts on coastal aquifers, vegetation and halieutic resources. These are elaborated below.

LAND LOSSES DUE TO COASTAL EROSION AND INUNDATION. Land losses due to erosion were determined by applying the modified Bruun rule (Nicholls *et al.*, 1995). For land losses due to inundation the formula proposed by Hoozemans *et al.* (1993) was used. Three inundation levels were defined: minimum inundation levels by 2050 and 2100 and a maximum inundation level by 2100. The results are presented in Table 2.16.

The beaches found to be most vulnerable to SLR are those located south of Dakar where the sand availability is limited. This confirms the conclusions of the previous vulnerability and adaptation study (Dennis *et al.*, 1995). Small towns as well as fisheries and industrial infrastructures densely occupy this part of the coast. The Sangomar sand spit seems particularly vulnerable since it could disappear by 2100 with the maximum inundation level scenario. Since the sand spit is protecting the whole estuary, its disappearance will have major consequences for systems in the estuary.

Inundation in the Saloum estuary could include 27% of the land and about 50% of the mangrove area by 2050. With higher inundation levels, the entire deltaic plain could be inundated leading to the estuary being restricted to islands corresponding to old barrier islands that are the highest parts of the estuary. Major communication problems will arise in this situation since important portions of roads will be flooded. In the Cap Vert peninsula, areas that are likely to be inundated are much more limited but they are places of dense human occupation. Hence, the socio-economic consequences will be significant.

MODIFICATIONS IN COASTAL AQUIFERS. Coastal aquifers, which form a major source of fresh water all along the coast, will be affected by any change in precipitation, reducing the recharge, and by SLR, which will increase saltwater intrusion. This situation will worsen with the expected increase in population.

From the expert judgement and based on a climate change scenario (diminution of precipitation between 1% and 10%; rise in temperatures comprised between 1°C and 3.5°C and a 0.5 m SLR by 2050), it appears that the piezometric level will be lowered and that the recharge will decrease while saltwater will progress inland. These changes will have important impacts on the *niayes* since their location depends on the piezometric level. It is expected that some will disappear and others will be created. Along the coastal lakes, the freshwater reservoirs will be reduced. In the Saloum estuary, the saltwater intrusion will be increased by the salinization of the river.

The results obtained on the Dakar aquifer with the FEFLOW model, using a 0.5 m SLR and a 10% reduction in the natural recharge, show a net progression of the saltwater interface that will contaminate part of the harnessing field used for freshwater supply of Dakar (Faye *et al.*, 2001).

Table 2.16. Land losses (km^2; % of total beaches in the area in parentheses) due to coastal erosion and inundation.

	Year	
	2050	2100
Losses due to coastal erosion		
Cap Vert peninsula	0.24–1.79 (3.8–28.5)	0.77–3.95 (12.2–62.8)
Saloum estuary	0.07–1.82 (4–109)	0.19–4.02 (11.4–241)
Losses due to inundation		
Cap Vert peninsula	48–57 (3–3.5)	397.7 (25)
Saloum estuary	896–1690 (27–52)	2911 (89)

Also it must be noted that the increase in water demand due to population growth was not considered in this analysis. Thus, the water deficit is likely to be even larger.

IMPACTS ON COASTAL VEGETATION. Two types of vegetation were considered: the coastal vegetation along the north coast and the mangrove in the Saloum estuary.

Based on a comparison with the evolution of the flora with drought (Hubert, 1917; Trochain, 1940; Raynal, 1963; Michel et al., 1969) it can be expected that along the north coast species not adapted to a rise of temperature (mainly the hydrophilic flora and in particular the relictual guinean flora) will disappear and the flora occupying the humid environments (lake sides for example) will be replaced by flora adapted to drought, salinity (halophytes) and new soil textures (more sandy due to the increase in aeolian erosion).

The mangroves present in the Saloum estuary are already close to their physiological limits especially for salinity (Diop and Ba, 1993). It is considered that temperature and rainfall changes will probably affect the phenology of the species. SLR will act indirectly, mainly by the processes of erosion and sedimentation. The recent breaching of the Sangomar sand spit indicates that changes in soil texture (becoming more sandy) due to massive influx of sands coming from coastal erosion of the spit have much more impact on the mangrove because they impede its reproductive cycle (seeds cannot germinate in sandy soils). An increase in salinity could also disadvantage the mangrove trees. Moreover, it is considered that with a 0.2–0.5 m SLR, the mangrove could slowly migrate inland but that a rise of 0.86 m will have more important consequences with for example changes in the zone of the mangroves and difficulties in migration.

IMPACTS ON HALIEUTIC RESOURCES. In the open sea, the main consequence of oceanic warming will be a decrease in the intensity of the coastal upwellings (Mitchell, 1988; Tsyban et al., 1990). This is confirmed by palaeoceanographic studies (Lapenis et al., 1990) as well as by recent records of inter-annual variability in the intensity of upwellings linked with changes in oceanic temperatures (Oudot and Roy, 1991). This will induce reduction in the productivity of oceanic waters that is the major support for fisheries. This, combined with the potential disappearance of some mangroves and oceanic warming, will affect the halieutic resources in line with the observations made in years with less intense upwellings and low halieutic productivity (Oudot and Roy, 1991). In fact, the main resources on which fisheries are based like tuna, groupers and small pelagics will be affected inducing changes in species composition with, for example, apparition of species which prefer warmer waters like *Balistes carolinensis*.

In the estuaries, halieutic resources will be mainly affected by the destruction of mangroves, which in the first instance will enrich the estuaries but over time will reduce primary production. These mangroves are used as nursery and breeding grounds for a number of species, in particular shrimp, and by other pelagic resources which use mangroves in one or another part of their life cycle.

Main socio-economic impacts

For each of the case studies, the population and the economic value at risk were assessed considering the different inundation levels and socio-economic scenarios as well as discount rates (3% and 6%). The main results are given in Table 2.17.

The total population at risk in the two case studies, the people at risk in the estuary and the people at risk at the Cap Vert peninsula, varied between 180,000 and 16 million (between 1.2% and 12.4% of the total population of the Cap Vert peninsula alone). The population at risk estimates are more accurate than those found for the total Senegalese coastline during the first study (Dennis et al., 1995) mainly because in the first study population growth was not considered.

In terms of economic impacts, the main components at risk are private houses and agricultural production. The economic value at risk with a discount rate of 3% varies between 342 and 59,000 billion of CFA F (US$0.49 and 83.8 billion), which is also much higher than the values at risk determined in the first study.

Table 2.17. Socio-economic impacts of inundations and costs of adaptation.

	Year and inundation level		
	2050 minimum inundation	2100 minimum inundation	2100 maximum inundation
Population at risk			
Cap Vert peninsula	109,000	730,000	4,700,000
Saloum estuary	74,600	847,000	11,800,000
Economic value at risk with a 3% discount rate in billion CFA F (billion US$[a])			
Cap Vert peninsula	182 (0.26)	3,000 (4.33)	21,000 (29.32)
Saloum estuary	160 (0.23)	2,800 (3.94)	38,000 (54.49)
Costs of adaptation with a 3% discount rate in billion CFA F (in billion US$ [a])			
Cap Vert peninsula	8.3 (0.01)	2.9 (0.004)	17.8 (0.03)
Saloum estuary	170 (0.24)	40.6 (0.06)	41 (0.06)

[a]Exchange rate 1 US$ = 700 CFA F in 2001.

Moreover, the impacts of climate change on fisheries will be determined by the reduction in resources as well as by the increase in extreme events and coastal erosion which will affect some infrastructure. This in turn could induce increases in the price of fish – with consequences on food security, in particular for the poorest – and loss of revenues.

ADAPTATION OPTIONS. First of all two types of coastal protection were considered: sea dykes – mainly for urbanized and industrialized areas – and dune afforestation for coastal zones not yet densely occupied. In the two cases costs of adaptation are lower than the economic value at risk as can be seen in Table 2.17. This is a completely different result from the previous study that concluded the reverse, which can be explained by an underestimation of the economic value at risk (Dennis et al., 1995). The new results indicate that protection measures should be considered especially for the parts of the coasts that are particularly important in terms of infrastructure and population, such as towns and harbours.

Since these measures would not solve all of the consequences of climate change, other adaptation measures were considered such as an integrated coastal zone management plan, better management of all coastal and marine resources (water, flora, halieutic resources), rehabilitation of salted soils in the Saloum estuary, legislative and institutional arrangements and the creation of a research centre. However, no cost estimations were made for these options.

2.10.4 Experiences and lessons learned

This vulnerability and adaptation study was a very rich experience. It was the first time such a study was conducted by a Senegalese team and with a multidisciplinary approach. Most of the members considered that they learned a lot from each other during this work that lasted almost 3 years. This exchange of experience not only led to new skills but also to ways of thinking differently than from a specialized approach. Regular meetings of the team contributed greatly to this and exchanges were very fruitful for all the members. We had the chance to discuss our ideas and approaches to define a common approach.

Of course, this study had its own limitations. We had to rely heavily on expert judgement mainly because some models were not available (i.e. for vegetation evolution) and data availability was limited. The models used were the Bruun rule and the more complex FEFLOW model. These were adapted to the specific circumstances for the case study by an expert who had the necessary data and experience to run the model. Another limitation was the fact that tourism was not considered in this study. Moreover, the qualitative nature of most of the evaluations limited the scope of the identification of economic impacts on, for instance, fisheries and tourism. Hence the economic evaluations should be considered as underestimations.

It was also the first time that this kind of study benefited from public communication.

Of course, it was limited to a workshop where major stakeholders were represented but the media gave much attention to the subject. This is interesting since at the beginning of the IPCC work and UNFCCC negotiations the feeling was that climate change was somehow too difficult to explain to the general public. Certainly this work helped to improve the way decision makers and the public are informed on climate change issues. This can partly explain why, when a few years later, a former Minister of the Environment explained climate change to an almost illiterate audience it was surprising to see how people clearly understood the key issues. The public awareness was also enhanced by new extreme climate events, such as the cold and rainy event that affected the northern part of Senegal in January 2002. This event was completely unexpected and brought about serious damage to private houses, agricultural land and livestock.

2.10.5 Follow-up research

Unfortunately, the study was not followed by a new research project. This, combined with financial problems, induced a dispersion of the research team. From all the team members only two could continue research work on climate change issues. One researcher is actually engaged in thesis work on the modelling of the Saloum estuary aquifer including considerations of climate change. The other has started a vulnerability and adaptation study on the impacts of SLR on the Senegal Delta. The other experts went back to their former research fields.

It is also noticeable that the expert that developed the climate change scenario is actually the principal investigator of an AIACC (Assessments of Impacts and Adaptations to Climate Change) project responsible for the development of regional climate change scenarios for the Western Sahel.

However, the limitations indicated show that further research is needed in areas such as the integration of tourism and fisheries in the economic evaluation and the assessment of other methods to protect the shoreline (for example beach nourishment, which was not possible to examine due to lack of data). It is also important that some protection techniques that are currently not used in Senegal be tested before any large-scale implementation.

2.10.6 Policy implications

The study did not conclude with concrete policy measures with the exception of the implementation strategy where coastal protection and management is suggested as a short-term option. A first order project proposal to test coastal protection methods is given in the first National Communication of Senegal (République de Senegal, 1999).

Senegal lacks the funds to develop new adaptation study projects. It is considered that financial resources for further projects must come mainly from external sources, with the national budget being used for other urgent needs.

It is expected that the second phase of the NCCSAP will allow concrete actions to be developed in order to go beyond these previous studies.

2.10.7 Conclusions

The vulnerability and adaptation studies conducted under the NCCSAP were a successful experience not only in terms of results but also for a better comprehension of potential impacts of climate change as well as for the multidisciplinary approach that was developed. Real progress was made in understanding the complexity of the vulnerability of our country to climate change.

The main biophysical impacts include land loss due to erosion and inundation, and salinization of surface water and aquifers. In the areas studied it was estimated that in 2050 over 180,000 people will be at risk and that the value at risk will amount to approximately half a billion US$. Adaptation options include coastal protection, dune afforestation and integrated coastal zone management. Costs of adaptation were estimated to be US$0.25 billion.

Main Publications From the Research

Niang-Diop, I., Dansokho, M., Diaw, A.T., Diouf, P.S., Faye, S., Gueye, K., Guisse, A., Ly, I., Matty, F., Ndiaye, P. and Sene, A. (2000) *Etude de Vulnérabilité des Côtes Sénégalaises aux Changements Climatiques.* NCCSAP/Ministère de l'Environnement et de la Protection de la Nature, Dakar, Senegal, 151 pp.

République du Senegal (1997) *Com-munication Initiale du Senegal a la Convention-Cadre des Nations-Unies Sur les Changements Climatiques (CCNUCC).* Ministère de l'Environnement, Dakar, 118 pp. Available from: http://unfccc.int

République du Senegal (1999) *Stratégie Nationale de Mise en Œuvre (SNMO) de la Convention Cadre des Nations Unies sur les Changements Climatiques (CCNUCC).* Ministère de l'Environnement, Dakar, 53 pp. Available from: http://unfccc.int

2.11 Suriname

C. Becker[19]

2.11.1 Introduction

With the signing of the Rio Convention in 1992 Suriname became the 112th party to the United Nations Framework Convention on Climate Change (UNFCCC), with a commitment to meet international obligations through presenting an inventory on greenhouse gases (GHGs) and impacts of climate change to the Convention. Suriname welcomed the financial and technical support of the Netherlands government to start the project 'Country Study Climate Change Suriname (CSCCS)'. This project started in 1997, prior to the ratification of the Convention by the Suriname Government, and finished in 2000. Two main objectives were formulated for this project: the first objective of the CSCCS was to produce a first national inventory of GHGs, determined by the obligations following from signing the Convention, and the second objective was to map the vulnerability of the coastal zone to sea level rise (SLR). Both studies were carried out for the base year 1994. Emphasis was on the second part of the project since many natural and socio-economic systems were expected to be disrupted by SLR.

In recent decades Suriname experienced an ongoing attack from the Atlantic Ocean over the largest part of her coastline, which, except for a few kilometres, is protected by mangrove forests. At certain locations along the shoreline, e.g. the coast of Coronie district, erosion has reached dramatic levels causing large-scale inundation and sudden changes in local ecosystems. This resulted in degradation of the environment for fish, shrimps, birds, turtles and other species in the coastal area. In addition, expanding areas of agricultural land and urban areas are affected by degradation of the coast. With the main population (±90%) living in the flat and low-lying coastal zone Suriname is highly vulnerable to climate change impacts due to SLR.

The studies contributed to capacity building on climate change issues and raised the awareness of policymakers and the public of the threats of climate change, in particular that of SLR. Moreover, the study results emphasized the need to implement Integrated Coastal Zone Management (ICZM). In addition, policymakers were advised to initiate policies and measures to cope with climate change problems, for instance to establish necessary institutions and support systems for short- and long-term policy development.

2.11.2 Approach

The Intergovernmental Panel on Climate Change (IPCC) 'Common Methodology' (Carter et al., 1994) was used as the basic approach in the study. The project started with preparations of an inception report and detailing the methodology for Suriname. For the description of the impacts of SLR the coastal profile was determined based on existing data and some field measurements. In order to calculate the risk of

---

[19] Meteorological Service, Magnesiumstraat 41, Paramaribo, Suriname. Tel.: +597 491143, Fax: +597 490627, Email: cbecker@sr.net

SLR and to indicate the regions at risk, a future coastal profile was determined. Because of the limited documentation on the future development of the country, the framework of the future profile was based on generally accepted theories and expert judgement. A Geographic Information System (GIS) based flooding and flood risk model was used for the vulnerability and impact assessment.

To analyse the vulnerability and impacts in more detail a pilot study was scheduled within the project. One of the goals of this pilot study was to train the staff in coastal zone management issues with respect to climate change. The pilot study also analysed promising adaptation measures. The results of the flood risk model were discussed in various workshops with different stakeholders such as scientific experts, policymakers, NGOs and others. The last national workshop in the project also identified and discussed a possible follow-up project with respect to climate change and coastal management.

2.11.3 Results of the coastal zone management study

At the start of the study, little knowledge on the climate change issue and its possible impacts was available in the country. The concept of SLR and its consequences for coastal areas were unknown to many policymakers and scientists. Only limited, 'hard to reach' data, mostly not digitized, were available. In particular, topographical information was lacking. Some information was known to exist, but it would take a special project to collect and digitize the data. Because climate change issues were unknown to many government officials, it took much time and effort to explain and convince them about the necessity of the project and its social relevancy at the national as well as international level. Their cooperation was needed to get information on and access to existing data. In many cases, this was successful. Box 2.5 presents details on the data collection phase.

Because of the limited time to finish the project and the problems of access to data in the early stages, it was necessary to use mathematical interpolation methods to fill the data gaps. This was especially the case with hydrological data, including water levels and saltwater intrusion. For other sectors such as geomorphology, ecology, and socio-economy, this was not possible. Some data were derived from the stakeholder meetings and sessions with third parties. Besides expert judgement (using similarities of more or less identical

---

Box 2.5. Data collection for ICZM.

Prior to the CSCCS, necessary data for assessing the vulnerability of the coast were scattered, not only over the various government departments and ministries, but also among the various NGOs and consultancies within the country. It was therefore a huge and difficult task to collect all these data, which, given the time and financial limitations, may be regarded as a successful event. The process of data collection led to many questions dealing with methods of data collection, rehabilitation of old observation networks, establishing new observation networks, data processing and establishment of a central database. Data were collected on the following four categories or sectors: (i) ecology; (ii) geomorphology; (iii) socio-economy; and (iv) water resources. From these four groups, the ecology sector provided most data, followed by geomorphology, water resources and socio-economy. The following sources for data collection were used: (i) existing publications; (ii) unpublished data collections; (iii) maps; and (iv) verifiable indications and experiences of local people and experts. Despite all these efforts large parts of the study area remained unknown, owing to, among other reasons, the inaccessibility of the area. Necessary information about these areas was then derived from satellite images or from interpolations and expert judgements. However, more and precise information about the sectors mentioned above remained a (pre)requisite until the project ended. Notwithstanding these shortcomings and gaps, the data collection gave, after its validation, a reasonable and well accepted representation of the coastal profile, on which basis the vulnerability has been determined. All the stakeholders, including the government, scientists and NGOs, now stress the importance of a database centre.

cases in the region or neighbouring countries) information from local experts and individuals was used to complete the socio-economic picture. Field trips and awareness meetings also contributed to this. Gaps were filled by indirect interpretation of the available data. In some cases, expert judgement was applied. This was especially the case with determination of the hydraulic conditions zones and the protection sections of individual areas in the coastal zone. The lack of topographic data and detailed information about the existing infrastructure, which were the main bottlenecks for determination of these zones, was solved by interpretation of vegetation maps, soil maps, knowledge from the local people, scientists involved in these areas and expert judgement. These data were mapped carefully through application of GIS.

GIS analysis results provided the necessary information for the Flooding and Flood Risk model. Calculations were made for: (i) the present situation; and (ii) the future conditions. Both cases were subdivided in (a) SLR = 0 m and no development and (b) SLR = 1 m and development. Of these two cases scenario (b) became a difficult task to solve. This is particularly the case with the determination of the future developments of the various natural and socio-economic systems and land-use categories within the coastal zone. Since adequate future planning and the necessary guidelines were missing for the determination of future developments, it became quite difficult for the team leaders to value these developments in the future and the impacts of their interrelations. Of particular importance were the assessments of the socio-economic developments, which basically are determined by the government in power. However, careful analysis of the present coastal profiles and intensive brainstorming processes with consideration of the possible climate change, has resulted in an acceptable outlook of the future development scenario.

In the course of the project, a few training sessions and training workshops were organised to make the local experts acquainted with this matter and to raise awareness. Experts made available by the Dutch counterpart conducted the sessions.

The first problem arose when setting the boundary conditions of the study area. There were no detailed topographical maps of the coastal area, except for parts of the Paramaribo, Wanica and Nickerie districts and of the east–west road connection along the coastal line. The lack of data made it difficult to determine the southern border of the area. From hydraulic and hydrological points of view, the southern border of the study area should be fixed based on the tidal effects, which under these circumstances reach up to the first rapids (waterfalls) in the main rivers of Suriname. Hence, the southern border of the research area was finally agreed near the first rapids/waterfalls in the main rivers. For the remaining area it was decided that the contour lines necessary for the flood risk assessment were to be derived from soil and vegetation maps.

With regard to the future development (projections) in the country, only the 'long-term development plan' published by the Planning Bureau existed. It was assumed that the overall population growth would be minimal, about 1.2%/year, and that Paramaribo, the capital of Suriname, would expand the most, by about 3%. In addition, the economic activities and infrastructure would grow with the expansion of the capital and its suburbs. Furthermore, it was assumed that some grass swamps and mangrove forests will be converted into rice fields and aquaculture. Banana plantations were assumed to remain unchanged. Box 2.6 presents a storyline on the importance of mangrove forests for Suriname.

In reality the general development trends in the coming (30) years will depend highly on, among others, government regulations. The policy of the government of Suriname is directed towards the development of industries. The main emphasis in these developments will be on bauxite, crude oil and gold production. Development of the agricultural sector will need large investments, while other sectors such as horticulture would need to be reorganized substantially.

Erosion and sedimentation processes at the coastal line are, among others, determined by sediment transportation from the Amazon region. How global climate change will affect this phenomenon is unclear. Moreover,

> **Box 2.6.** Importance of mangrove forests.
>
> Large parts of the Surinamese coast are occupied by mangrove forests. These forests are important natural ecosystems in the coastal zone and provide a natural coastal defence system. During recent decades erosion at certain locations along the coast, especially in the Coronie (site Totness) and Wanica (site Weg naar Zee) districts, has increased and taken dramatic forms. From the study it appears that at higher sea levels and changing wind patterns (wind velocity and direction) the level of erosion will increase further and losses will increase drastically. These two locations are typical examples of sites where the mangrove forests have been cleared for agricultural purposes and settlements and infrastructure were established. This has led to adverse changes in the sedimentation pattern in the coastal zone. To understand the complexity of the coastal dynamics, an in-depth and detailed study of the coast, and in particular the mentioned sites, is needed. The government of Suriname is very interested in the defence of the coastal zone, since the ongoing and persistent erosion along the coastal zone causes huge damage to the natural ecosystems as well as to the local infrastructure and settlements on an annual basis. Hence, it wants to use the preliminary outcomes of the study as inputs for policy regarding the coastal protection and coastal zone management.

changes in the sea current pattern are unknown. It was assumed that no changes in sedimentation transportation will occur and that (mud) bank formation will continue at higher sea levels. Furthermore, it was also assumed that waves, due to changes in wind patterns, and oceanic current regimes will become stronger and accretion will continue. However, this will lead to increased erosion.

The natural defences such as mangroves and shell cheniers are likely to protect the coast, except for those areas where intense erosion has destroyed them. Here tidal waters will penetrate into freshwater swamps and coastal aquifers. It was assessed that the mangroves remain landward and occupy the extended saline intertidal area under SLR. Moreover, with the increase of seawater levels engineering structures in Nickerie and Coronie districts can be expected to be less effective in their function to prevent flooding. The combined effect of all these factors will seriously affect, among others, the harbour facilities in the country. Consistent with existing climate data of the area for the past 100 years and General Circulation Models (GCMs), the study also assumed a decrease in precipitation in the future and thus the freshwater resources.

On the basis of the above mentioned assumptions (gradual SLR, conversion of mangroves into agricultural land, ongoing mining of shells and sand in the coastal area, expansion of the urban area and other infrastructure related to for example oil extraction, pollution of coastal waters by municipal and agricultural lands) it was found that the coastal ecosystems are highly vulnerable to future developments and to climate change. Under these circumstances, it was assessed that the country will suffer large-scale losses of goods and services of the estuarine zone, which also function as the buffer zone for saltwater intrusion, sedimentation, wave attacks, etc. To preserve these ecosystems the number of protected areas needs to increase and the total area covered needs to be enlarged.

The results of the CSCCS project showed the vulnerable state of the coast. This: (i) raised the awareness of the stakeholders; and (ii) urged the government to take actions against the increasing effects of the SLR. The Suriname coast is, except for 7 km out of 386 km, not protected from the attacks of the sea. Changes in the climate patterns, e.g. wind speed and wind direction, together with SLR and other developments in the coastal area, form a great threat to the existence of the natural and socio-economic systems on the coast. Moreover, the government is at present unable to cope with these challenges and problems, and consequently flooding, erosion and salinity intrusion are expected to keep pace with climate change. There is a need for ICZM in Suriname. ICZM is also a powerful tool for raising awareness. The study identified several short-, medium- and long-term actions that could be taken.

The short-term actions include:

- Start with an awareness programme (radio, television and newspapers).
- Establish a database for wider use by stakeholders (national and international).
- Organize regional workshops.
- Develop and incorporate training in the curriculum of the university programme.
- Exchange of knowledge through missions by experts from countries with similar problems and from countries that have experience in this field.

Medium- and long-term actions include the establishment of a Coastal Zone Management Authority with the following objectives:

- Enhance the understanding of the mechanisms of land and sea interactions in the coastal area. Develop and establish information systems in the coastal area for solving possible resource conflicts.
- Develop and set up an ICZM plan including coordination and management of the coastal zone. This could be done through training of local staff and through development of ICZM tools and concepts.
- Strengthen and establish cooperation between Suriname and neighbouring countries and between Suriname and the Netherlands.
- Establish a portal system.

2.11.4 Experiences and lessons learned

The CSCCS included a considerable quantity of human resources and materials of institutions and government departments, including the University of Suriname. The study contributed to an increase in the scientific and technical capacity within these organizations.

The realization (execution) of the study was a rapid learning process, which implied 'learning by doing'. Much information was known to exist, though widely scattered in the government institutions and departments. The study benefited from guidance by Dutch specialists, consultants and experts, which fostered human resources and technical capacity in the country to a significant degree. National specialists can now perform follow-up studies with less technical support from international experts.

Furthermore, the study also encouraged the establishment of the present National Institute for Environmental Issues in Suriname (NIMOS) where other conventions are also managed. In the course of the project, implementation scientists from different disciplines worked together to produce the desired results, which are published in various papers, reports and maps.

This process proved that national scientists were able to contribute to multidisciplinary workgroups that can develop knowledge and reproduce activities. Hence, the project delivered a new generation of planners and analysts. The new generation of scientists has learned to recognize and understand the complexity of the coast in the perspective of climate change, identify problems and where possible offer proper options for solutions. An example of how scientists working for the CSCCS have done this for the disrupted drainage systems in the coastal area is provided in Box 2.7.

An important and time-consuming factor following from the lack of knowledge on climate change issues was the divergent views offered by the participants on subjects such as the scale of inundation of the coastal area, coastal processes like erosion and sedimentation under future conditions, and possible future socio-economic developments. This often resulted in long discussions, and unwillingness to accept each other's view. Many of these views and proposals were based on difficult to verify or even insufficient data. Paucity of law and regulations also formed a burden in the execution of the study. Although the political situation was stable during the study implementation phase, it was difficult to calculate the development of the future socio-economical processes due to the lack of strong political impulses.

The lack of experience in this field and the huge amount of data that had to be compiled in this phase and the short time frame, especially for the production of the thematic maps, placed a heavy burden on the project. Moreover, this particular study was new for

> **Box 2.7.** Disruption of the drainage systems in the coastal area.
>
> The low and flat coastal areas of Suriname form the most fertile zone of the country, and hence provide necessary conditions for large-scale agricultural practices, residence and for other industrial developments in the area. One common condition required for all activities in the area is adequate drainage. Drainage of this area has been taking place naturally, through rivers, creeks and wetlands, and artificially, through sluices and culverts. In general, drainage of the country is based on gravity flow, however, as the sea level rises, drainage of agricultural land, rural and urban areas will stagnate and the threat of inundation, combined with heavy rainfalls, which are observed more frequently, will increase. Moreover, due to the maintenance in arrears of the hydraulic constructions (sluices, weirs, dams, culverts, etc.) saltwater intrusion takes place continually, resulting in shrinking freshwater resources in the coastal area, and in particular, in the rural areas. In the urban areas, in particular Paramaribo, additional hydraulic pumps have been installed to combat the problem of inundation. These developments in the drainage systems are thanks to the CSCCS project, which for the first time mapped the vulnerability of these areas to global climate change.

many scientists and therefore it was difficult to assess the impacts of SLR over a period of almost 100 years (in 2100), especially on the development of socio-economic scenarios until 2100. These conditions were simplified by assuming the sea level to rise by 1 m not in 2100, but in the year 2025. This assumption brought a lot of clarification for many of the participants, but was again confusing for others. These confusions and improper understanding of the conditions of the project led to many revisions of, among others, the maps needed for the assessment and particularly the economic values of different land-use categories. Many land-use categories, such as swamps, lagoons, ponds, grass swamps and mangrove forests, were valued for the first time in this study. Since there was a lack of necessary expertise and information, proper evaluation of these land-use categories was difficult and caused lengthy discussions on their economic value. In most cases, the valuations were finally based on expert judgement. According to some experts, these values tended to be underestimated.

The scheduled pilot project that was supposed to train the local staff on the job could not be executed due to financial and time constraints. The financial support system of the project caused a number of problems during execution of the project.

Other lessons drawn from this study are:

- Many stakeholders were unable to consider a time frame for the study of 100 years. It became clear that for studies to appeal and be relevant to policymakers and the public shorter time frames needed to be assessed.
- Results of and recommendations from the study were in some cases misquoted by users. Recommendations were taken as assignments and not as guidelines towards the mitigation of the severe impacts of SLR. This showed the importance of comprehensive reporting of results.
- More and frequent dissemination of information should take place, especially outreach activities.
- Many people are unaware of the major consequences and losses due to climate change despite the fact that a survey showed that they had observed negative changes in their environmental and economic conditions. An intensive awareness programme was missing in the project. Awareness was only created in a relatively small stakeholder group.
- Adequate data were lacking, which delayed the study and forced the researchers to make numerous assumptions.
- The study showed that the country of Suriname is not ready nor able with its existing expertise, technology and financial means to cope with the problems of SLR. Foreign support is needed for the country to cope with SLR and design and implement ICZM. Furthermore, the rights and responsibilities of individuals and public agencies are not clearly specified, leading to possible conflicting situations under the current legis-

lation and regulations. In certain areas within the coastal zone several different stakeholders practise economic activities, which in most cases are not regulated. Here, the current regulations are obsolete or missing and need to be adapted or elaborated. A recent example was the drainage of polluted water from the agricultural lands into the coastal wetlands, negatively affecting the unique ecosystem. Other economic activities, such as oil exploration, in the coastal area intervene not only in the proper functioning of the ecosystems, but also affect other small-scale economic activities, such as beekeeping, swamp fishery, hunting, etc. This may cause conflict situations among the stakeholders.
- With SLR the poor are the most likely to suffer from inundation, flooding and other negative impacts such as salinization and loss of land.
- No assessment has been made of the poor, but from observations, especially in the Coronie district, it follows that many of the inhabitants live in the old abandoned plantations, close to the ocean. In the other districts, such as Commewijne and Nickerie, this is also common. These people are not able to protect themselves from the threat coming from the ocean. By contrast, the people living in the north of Paramaribo, also bordering the ocean, are more prosperous and therefore less vulnerable.

The vulnerability was evaluated by calculating the probability of extreme hydrological events of a certain intensity (e.g. inundation level and duration) in combination with 1 m SLR. It was quantified in terms of potential damage to persons or objects ('elements at loss and or at risk') in the regions at risk. By producing risk maps, using GIS, and surveys of vulnerability and hazard maps, areas vulnerable to SLR were identified. These maps indicated areas where urgent action is needed to improve the natural coastal defence systems. Much time was spent in discussing issues from different professional views. This process has proved to be necessary to reach consensus among the stakeholders.

The next step was to interpret the vulnerability and adaptation assessment for each sector and to review existing policies. This was even more difficult, because such policies were not clear and were partly lacking.

The IPCC Seven Steps Method proved beneficial for determination of the vulnerability of the coast, however, the time frame for implementing the different steps was too short. Moreover, the relevancy of each step needed to be emphasized and explained by the experts.

Adaptation measures are not fully worked out in the study. This is simply due to the lack of required data. There is a need to cover gaps in data and data management in the country. This highlighted the need for the establishment of a data bank, which should provide a useful basis for determining adaptation options, and regional coordination, especially regarding satellite images, rainfall data and migration problems. Establishment of a data collection network is required, together with appropriate observations and data gathering techniques, and quality control. Baseline data analysis should also be included, where national and regional experts are involved.

The use of more local knowledge across the entire country is considered necessary. Training should be enhanced and more means should be freed for these purposes.

Use of methodologies

As mentioned before, the IPCC Seven Steps Method was used to determine the vulnerability of the coast. The Dutch consultants explained the method in one workshop session for the national project managers, team members and technicians. The local team decided to use data derived from the existing studies, maps and reports. These data were mainly spots and studies on a local scale, which were often limited to a single item or sector.

An overall picture of the coast was not available. Therefore, satellite images and other data were purchased from which necessary maps were produced for sediment, soil, catchments, vegetation, population, geology and rainfall. The respective consultants and scientists working on the project needed to update these maps, because for some areas no recent satellite pictures were available. Additional data for updating these maps were received from various sources, e.g. from government departments, consultants, institutions, NGOs and individuals. In many cases, simple assumptions or interpolation of events were used to fill gaps in the data. It was the first time that huge data sets were presented in an integral manner using GIS. Investment in creating data and training personnel in the use of GIS and other applications was essential for the success of the project.

2.11.5 Follow-up research

Before closure of the CSCCS project in 2000, Suriname started another project with the help of the United Nations Development Programme (UNDP)/Global Environment Facility (GEF), aiming to produce Suriname's first National Communication to the UNFCCC. It was agreed that the results of the CSCCS would be used as the basis for the National Communication. However, the National Communication is not finished yet.

In the recommendations of the CSCCS, a follow-up study was proposed, but the economic situation of the country prohibited the execution of even part of the recommendations. The donor of the first NCCSAP project financed a follow-up project that started in July 2003 under the responsibility of the Ministry of Labour, Technology and Environment. The project will focus on the area of adaptation with specific emphasis on linkages between poverty and climate change.

2.11.6 Policy implications

With the signing of the Kyoto Protocol, the government of Suriname not only committed itself to report to the UNFCCC, but also to undertake the necessary steps to 'cope with and adapt to potential climate change and sea level rise, including the development of globally acceptable methodologies for coastal vulnerability assessment, modelling and response strategies'.

Owing to political priorities, the change of government and the financial situation of the country, among others, the findings and recommendations of the study were not published until 2003. Box 2.8 presents some highlights of the effects of the recommendations from the study.

The choice of the study was made as a result of consultations with the Ministry of Planning (national planning office) while the project was monitored by this ministry through one member in the project steering committee. The project members kept their respective ministers informed of the progress of the project and the problems in the field. The committee took the feedback from the stakeholders into account. The results from the study were used in long-term planning of the sea defence in the north-western part of the country. To protect the capital investment in the low-lying rice area of the Nickerie district, the height of the sea fence was increased in line with the expected SLR and the feasibility of the adaptation.

Paramaribo, with a population of about 200,000 inhabitants, lies less than 15 km

---

Box 2.8. Effect of the recommendations for society, including the policymakers.

The study clearly showed that Suriname's vulnerability to climate change can be ranked among those of the small island states. These results shocked the society, including the government. Many of the inhabitants were not aware of the ongoing problems in the coastal area and the need for a better management of the resources. With the publication of these results a milestone had been reached, which contributed to the preparation of coastal zone management plans for the northern parts of the districts bordering the sea, preparation of a 'discussion paper' on marine and coastal management for the development of a framework policy and a strategic plan for sustainable management urged by the government, and new proposals for detailed studies of the physical processes that affect the coastline. In addition, government officials (e.g. ministers) are using the results and conclusions to promote their policies regarding protection and sustaining the development of the coast.

from the coast. The present drainage system, which is based on gravity, is becoming ineffective. SLR will worsen the situation. To adapt the situation, 'a master plan on urban drainage of greater Paramaribo' has been set up, indicating the need to install necessary pumps and means for artificial drainage.

Many government departments took advantage of the training and dissemination activities and the application of GIS. The Department of Natural Resources and Environmental Assessment (NARENA) of the University of Suriname particularly benefited from the project as they have broadened their know-how and capabilities, in particular in analysing and processing satellite images. Hence, the NCCSAP studies have impacted various disciplines in the country and promoted the use and application of GIS information on a broader scale.

2.11.7 Conclusions

Notwithstanding the complexity of the discussions mentioned earlier and the difficulties in executing the project, it is possible to draw some principal conclusions that are central to this project. These are:

- The use of the seven steps in determining the vulnerability of the coast has had good results. The methodology was sufficiently flexible to meet most of the needs.
- A wide variety of techniques and know-how was gained in assessing the impacts of climate change during the execution of the project. It can be concluded that the transfer of knowledge was successful.
- The need for integrated coastal zone management became obvious and necessary to halt certain negative processes and affect ongoing development in a positive way.
- The need for continuing and furthering the study of coastal vulnerability was highlighted, including the need for coordination of the different activities in the coastal sectors and the policies of the different government departments concerned.
- The project can be regarded as a powerful awareness and education programme for the nation, since this was the first attempt in determining the indirect impacts of climate change for Suriname.
- In this regard some strategies for mitigating the impacts of the SLR have been proposed.
- A database has been set up, which may be regarded as an important result.
- Implementation will be not an easy task. Stakeholders and agencies must coordinate and harmonize their policies and programmes. Policymakers should have the political will to put the measures in place and provide the necessary resources.

Main Publication From the Research

Country Study Team Climate Change (1999) *Country Study Climate Change Suriname and First Steps towards Integrated Coastal Zone Management.* 83 pp. with eight Technical Reports: Geomorphology. Profile. 29 pp. (Dixit, P.C.), Geomorphology. Prediction. 12 pp. (Dixit, P.C.); Socio-economics. Profile. 24 pp. (Consen, J.R., Sanchit, R. and Tawjoeram, J.), Socio-economics. Prediction. 29 pp. (Consen, J.R., Sanchit, R. and Tawjoeram, J.); Ecology. Profile. 28 pp. (Baal, F., Hiwat, M. and van der Lugt, F.), Ecology. Prediction. 23 pp. (Baal, F., Hiwat, M. and van der Lugt, F.); Water Resources. Profile. 77 pp. (Amatali, M. and Naipal, S.), Water Resources. Prediction. 24 pp. + Appendix (5 pp.). (Amatali, M. and Naipal, S.); Flood and Flood Risk Modelling (Hoozemans, F.M.J.); Geographic Information System (van Veldhuizen, H.); GHG Inventory 1994. 108 pp. (Becker, C., Breinburg, H., MacDonald, H., Playfair, M. and Ramdihansing, H.); Final Report 1999 (Amatali, M., Becker, C., Hoozemans, F.M.J., Leenen, H., Naipal, S. and van Veldhuizen, H.).

2.12 Vietnam

Nguyen Ngoc Huan[20]

2.12.1 Introduction

Vietnam is a country with a coastline length of 3822 km (excluding islands) that runs from Mong Cai in the north to Ha Tien in the south. Vietnam has a long coastline and a narrow hinterland. Flooding in the coastal zone is mainly a result of high river discharges, elevated sea level during typhoons and weak dykes.

Vietnam is a signatory to the United Nations Framework Convention on Climate Change (UNFCCC) and is making preparations to meet its commitments. A global inventory made in preparation of the United Nations Conference on Environment and Development (UNCED) in Rio de Janeiro showed that Vietnam is one of the most vulnerable nations as regards the threat imposed by climate change, including sea level rise (SLR). It was clearly recognized that the vulnerability needs to be quantified and response strategies need to be developed. Vietnam was thus chosen to conduct three pilot studies on the impacts of climate change on the coastal zone.

These studies, titled 'Vietnam Coastal Zone Vulnerability Assessment and First Steps towards Integrated Coastal Zone Management (Vietnam VA Project)', were conducted in Vietnam from November 1994 to April 1996. The aim of the Vietnam VA Project was to assess the vulnerability of the whole coastal zone of Vietnam to increased SLR and related effects due to climate change. The first pilot study was a study on dyke erosion in Nam Ha Province (in July 1995), the second one was a study on flooding and lagoon management in Thua Thien Hue Province (in November 1995) and the third one was a study on planning in Vung Tau (in March 1996). The choice for the sectors was determined by their economic importance for the country as well as by the existing expertise in the country. The study areas are shown in Fig. 2.13.

In addition to the development of different national climate change scenarios, many studies have been developed such as the studies on recent historical setting, political developments, population, employment, subsistence farming, land use, economic growth and institutional arrangements for coastal zone management, etc.

The 1996 coastal zone VA study assessed the impacts of various scenarios for SLR – namely coastal erosion and inundation – on the entire coastline with rough estimations of population, land value and natural habitats and species at risk as well as an evaluation of the cost of adaptation, mainly protection. Within this context, the main objective of this pre-NCCSAP study on the impacts of climate change on Vietnam's coastline was to express a more comprehensive and integrated view of coastal vulnerability to climate change.

Assisted by the United Nations Institute for Training and Research (UNITAR) with its Climate Change Training Programme (CC:TRAIN), the government of Vietnam has set up a 'Country Team' comprised of representatives from key Ministries and operating via a coordinator, executives, a core team and four working groups. The Hydrometeorological Service (HMS) is a focal point in Vietnam for climate change issues and a leading member of the Country Team providing the core team secretariat and the Director General of HMS as the Coordinator and Chairman.

In addition to other activities initiated by the Country Team, recognizing the potential risks and the need to respond to the impacts of SLR on the coastal zone and its sustainable development, the Director General of HMS made a specific request for assistance with 'a study on vulnerability assessment along the guidelines as set up by the coastal zone management subgroup of IPCC [Intergovernmental Panel on Climate Change]' to the Netherlands Embassy in Bangkok.

Subsequent meetings and discussions in the Netherlands and Vietnam culminated in a pilot mission to Vietnam by coastal specialists from Poland and the Netherlands in October

---

[20] Centre for Application of Hydrometeorological Technology (HYMETEC) under the National Hydrometeorological Service of Vietnam belonging to the Ministry of Natural Resources and Environment, Email: hmec@fpt.vn

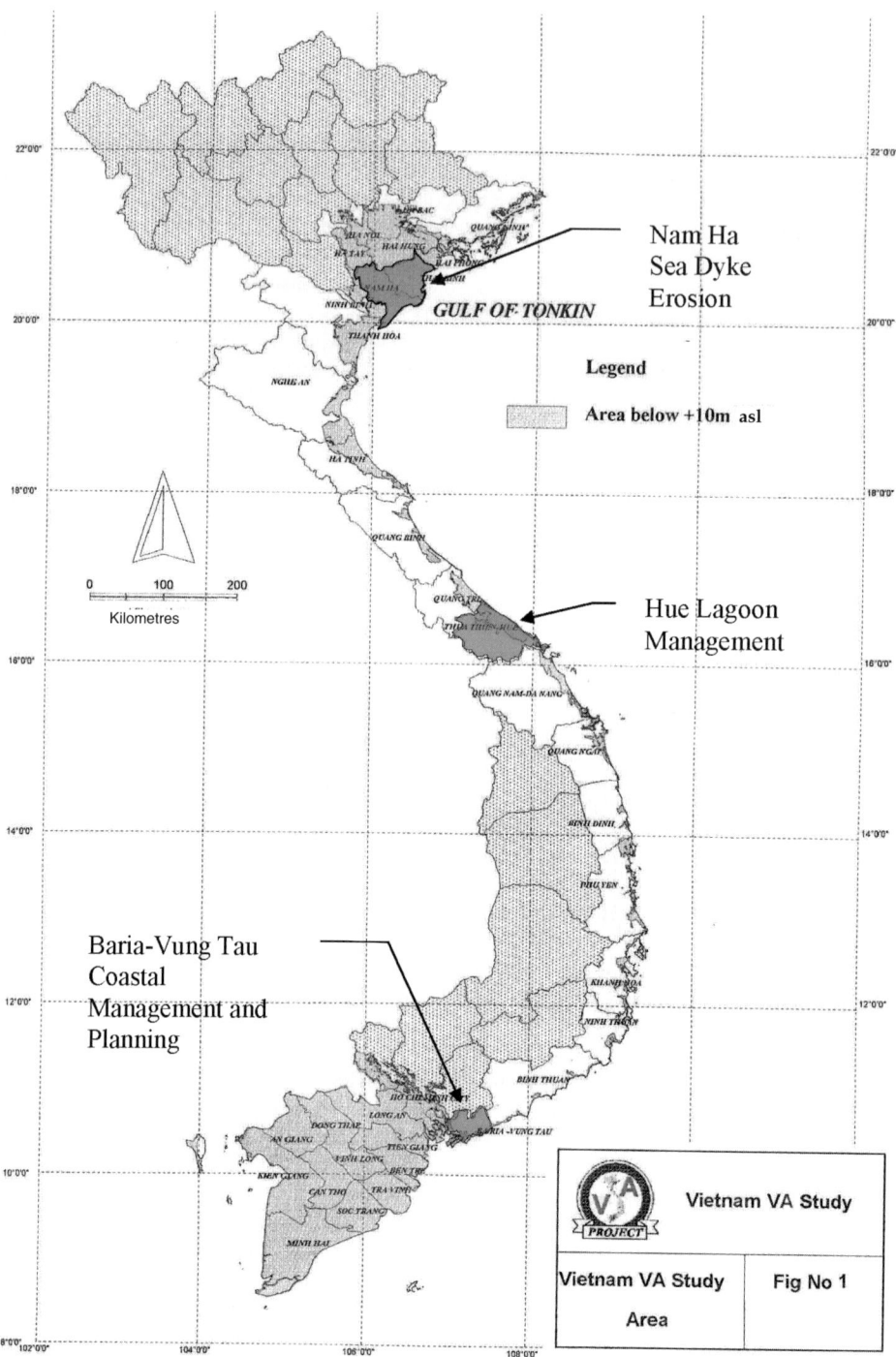

Fig. 2.13. The three study areas in Vietnam.

1993. As a result of this pilot mission an outline for an 18-month study in Vietnam was identified with the primary objective to provide the Vietnamese counterpart with assistance in executing a VA to assess the impacts of SLR on the coastal zone of Vietnam and by doing so strengthen the capacity of Vietnamese organizations to set up integrated coastal zone management. The project was initiated in November 1994.

2.12.2 Approach

Based on the results and experiences of the 1996 VA study, it was decided to include the following improvements in the NCCSAP-funded study:

1. Consider case studies that are representative of the main coastal environments. Three geographic locations were chosen: Nam Ha Province, Hai Hau District, where serious erosion of weak dyke defences is present; Hue in the centre of the country where a coastal lagoon system is present; and Vung Tau, where it is possible to take advantage of natural deep water port facilities and a burgeoning nearby offshore petroleum industry and where good recreational beaches and tourism are present.
2. Use a multidisciplinary approach by building a team constituted of personnel from the Department of Dyke Management and Flood Control (of the Ministry of Agriculture and Rural Development; in the study on sea dyke erosion in Nam Ha Province), personnel from the local Hue University, the Hanoi University and the Sub-Institute of Physics (Ho Chi Minh City; in the study on flooding and lagoon management in Thua Thien Hue Province) and local authorities or local specialists were consulted and informed (in the study on Vung Tau planning).
3. For each case study, consider not only the impacts of SLR but also other potential climate change impacts such as changes in rainfall, temperature and upwelling.
4. Perform a more comprehensive assessment of socio-economic impacts by using socio-economic scenarios and discount rates.
5. Have a broader approach of adaptation to enlarge the basic solutions developed by the IPCC/CZMS: retreat, accommodate, protect (IPCC/CZMS, 1991a).

The study used a mix of methodologies such as Geographic Information System (GIS) analysis to determine the areas flooded for four different scenarios with flooding: 1/year, 1/10 years, 1/100 years and 1/1000 years by using a SPANS EXPLORER software package. Flooding and Flood Risk (FFR) analysis combined the results from GIS analysis and the Geomanagement System (GMS) analysis, a data management system designed by Da Vinci Consultants in Belgium.

2.12.3 Results of the coastal zone study

The three case study areas are described in more detail in this section. Afterwards, the main results of the VA study are presented.

The case study areas

AREA 1: SEA DYKE EROSION IN NAM HA PROVINCE. This study area was limited by the Red River Delta coastal provinces of Thai Binh and Nam Ha, but the study focused on the serious erosion of weak sea defences in the Hai Hau district, Nam Ha Province.

Site visits were undertaken together with personnel from the Department of Dyke Management and Flood Control (of the Ministry of Agriculture and Rural Development) and in liaison with local authorities at provincial, district and commune level.

It became clear that the shoreline and foreshore (several kilometres seaward of the shoreline) were facing a serious sand depletion problem, unlike the majority of the Red River Delta coast that is accreting. Serious erosion of about 30 m/year is being experienced at the shoreline of the Hai Hau district. Former lines of sea defence were overtaken by the retreating coastline and are visible 200 m offshore of the present dyke-lined coast. The study area has eight mangrove species, 38 mollusc species, 30 crustacean species and 40 bird species. Economic activities related to the coastal zone include agriculture, aquaculture and fisheries. The area has a high potential for aquaculture in the Red River Delta – one of the most developed regions of the whole country.

AREA 2: THUA THIEN HUE PROVINCE. This study reviewed the problem of flooding and management in the coastal lagoon system around Hue in the central coast province of Thua Thien Hue. It was conducted in November 1995. Site visits were undertaken with personnel from Hue University, Hanoi University and the Sub-Institute of Physics (Ho Chi Minh City). Extensive discussions and meetings were held with local authorities, particularly with the People's Committee of Thua Thien Hue Province in the city of Hue.

The city of Hue has as major assets its cultural history (seat of power of the last emperor of Vietnam-Nguyen Dynasty), its extensive lagoon system (Tam Giang to Cau Hai Lagoon), its arable land, freshwater supplies and tourism resorts. In addition, aquaculture, port development and transportation are adding to the prosperity of Thua Thien Hue Province. However, serious threats jeopardize these assets, such as frequent (almost annual) extensive flooding from the 3000 mm of rainfall each year in the period September–November which can be accompanied by strong typhoons with storm surges. Salinity control problems threaten rice production (threatened by high salinity) and aquaculture (threatened by low salinity). Port development is hampered by sedimentation problems in some areas and erosion in others. Water levels will be affected but also salinity impacts and morphological changes can take place, with huge consequences.

The study reviewed the physical, environmental, economic and institutional properties and constraints of the region and set out a framework for addressing the complex issues in more detail.

AREA 3: VUNG TAU. This study reviewed the planning issues presently facing the area of Vung Tau in the south of Vietnam. It is well placed to provide both deeper port access for import of goods into the region (Ho Chi Minh City – Song Be – Vung Tau = 'Southern Focal Economic Zone') and to attract tourists to its sandy beaches. The region is almost completely flood free due to the absence of typhoons, which pass far north of Vung Tau, and the wetland ecosystems in its wide river mouths which prevent high flood levels near the coast.

In contrast, one of the largest remaining natural mangrove forests in Vietnam is endangered by port developments. The mangrove forest is not only an environmental resource but it also plays an important protective role for the local river trading ports and waterways leading to the Port of Saigon. Pollution owing to port development, management and protection of the beaches and dunes, tourism development in the coastal zone and industrial development of the petrochemical industry are all major issues that pose threats to the assets of the region.

An important inventory and visualization of the key issues was achieved which provides focus for careful consideration of planning and development options and a framework for further study.

Main biophysical impacts

AREAS AT RISK. Generally, the sea and estuary dykes are under-designed for the natural forces and processes. This results in high maintenance requirements or the acceptance of high damage and land loss. The highest costs for annual maintenance are incurred in Nam Ha (Area 1), where serious annual erosion and over topping of sea dykes regularly takes place. Table 2.18 presents an overview of the main impacts in the three areas investigated.

Table 2.18. Area lost and area at risk (km^2) in the case study areas in case of 1 m SLR.

	Coastal region		
	Nam Ha (Area 1)	Thua Thien Hue (Area 2)	Vung Tau (Area 3)
Area lost	0	0	0
Area at risk	1	1	9

THREATS TO WETLAND FORESTS. The present wetland forest area is only 30% of the wetland area in 1940. The creation of fish ponds was not only bad for the environment but it was also carried out ineffectively, resulting in the acidification of the ponds leading to low efficiencies and abandonment. In general, the tidal marshes are ecologically vital as they provide the spawning and nursery grounds for numerous bird, fish, prawn and mollusc species. A typical example is the Xuan Thuy reserve area at the mouth of Red River (Area 1).

THREATS TO COASTAL BIODIVERSITY. In general, the biodiversity in Vietnam is rapidly decreasing. This topic is of such national concern that Vietnamese scientists have recently published the *Sach Do Viet Nam* (*Vietnamese Red Book*; Ministry of Science, Technology and Environment of Vietnam, 1992), summarizing the status of threatened animals in the country. Table 2.19 presents the status of the major groups in terms of whether they are endangered, vulnerable, threatened, rare or undetermined species.

Another threat is the destruction and degradation of coastal vegetation and coral reefs by human activities such as marine tourism, port operations, and oil and gas extraction. These activities lead to coastline and dyke erosion and consequently to loss of critical shelters for fishermen. As mentioned earlier serious losses to mangrove areas are an additional threat.

SALINITY INTRUSION. Salinity intrusion in the coastal zone is increasing due to freshwater extraction for irrigation and drinking water and due to dam construction in some catchments. SLR will cause a higher penetration of saline water into rivers, creeks and streams as well as into groundwater systems.

MAN-INDUCED SUBSIDENCE DUE TO GROUNDWATER EXTRACTION IN VIETNAM.

1. Red River Delta. The highest subsidence of 176 mm from 1988 to 1992 (about 4 cm/year) was found in the south of Hanoi near Phap Van, while an average subsidence rate of about 10 mm/year was estimated for the greater Hanoi region. By 2010 the amount of groundwater exploited may increase to twice the present amount, while in 2020 a groundwater exploitation of about 1 million m^3/day is anticipated. Subsidence computations predict a total maximum subsidence of 750 mm near Phap Van under an exploitation rate of 1 million m^3/day.

2. Central Coastal Areas. The cities of Vinh, Hue, Da Nang and Nha Trang all use surface water for meeting their water demands. Only in Da Nang is some groundwater exploited on a small scale, while in Nha Trang wells are being used for pumping groundwater coming from rivers. Therefore, increased subsidence of these urban areas is negligible.

3. Ho Chi Minh City and Mekong Delta. Ho Chi Minh City and Vung Tau mainly use surface water for agricultural production and living conditions. Some wells are also being used for small groundwater exploitation of about 30,000 to 50,000 m^3/day, which is about 10% of the city's demand. Groundwater reserves in the Mekong Delta are very large. But its potential use is limited by three factors: salinity, permeability of aquifers and salinity intrusion during aquifer recharge. The safe yield of the basin has been assessed at roughly 1 million m^3/day. In general, the subsidence is not yet measurable, but it needs careful atten-

Table 2.19. Status of the major groups in terms of endangered, vulnerable, threatened, rare or undetermined species (Source: Ministry of Science, Technology and Environment of Vietnam, 1992).

	Invertebrates	Fish	Reptiles	Birds	Mammals	Total
Endangered	10	6	8	14	30	68
Vulnerable	24	24	19	6	23	96
Threatened	9	13	16	32	1	71
Rare	29	29	11	31	24	124
Undetermined	3	3				6
Total species in danger	75	75	54	83	78	365
Total species in the country	7,000	2,500	260	800	275	10,835

Table 2.20. Socio-economic impacts of inundation and costs of adaptation.

Impacts	Area		
	Nam Ha	Thua Thien Hue	Vung Tau
SLR 0 (no sea level rise)			
Population to be moved 1995 (persons)	0	0	0
Population at risk 1995 (persons)	629	42,020	3,454
Population to be moved 2025 (persons)	0	0	0
Population at risk 2025 (persons)	885	57,254	5,420
Capital value at loss 1995 (US$ mil.[a])	0	0	998
Capital value at risk 1995 (US$ mil.)	0	15	305
SLR 1 (sea level rise of 1 m)			
Population to be moved 1995 (persons)	0	383,605	58,463
Population at risk 1995 (persons)	8,142	37,462	955
Population to be moved 2025 (persons)	0	522,515	nk[b]
Population at risk 2025 (persons)	11,432	51,260	1,571
Capital value at loss 1995 (US$ mil.)	0	135	998
Capital value at risk 1995 (US$ mil.)	6	12	15
Capital value at loss 2025 (US$ mil.)	0	714	18,626
Capital value at risk 2025 (US$ mil.)	50	78	305

[a]Exchange rate 1 US$ = 15,500 VND (2003).
[b]nk, not known.

tion due to the area's sensitivity to the sea level.

Main socio-economic impacts

For each of the case studies, the population and the economic value at risk was assessed considering the different inundation and socio-economic scenarios. The main results are summarized in Table 2.20.

In terms of economic impacts, the main components at risk are private homes, agricultural and aquaculture production sites. Moreover, the impacts of climate change on aquaculture and fisheries will be determined by the reduction in resources as well as by an increase in extreme events and coastal erosion, which will affect infrastructure. This in turn could influence the price of fish with consequences on food security, in particular for the poorest, and loss of purchasing power.

Adaptation options

In the three case studies, information was extracted from the protection development plan drafted by the Vietnamese government Ministry of Agriculture and Rural Development (Department for Dyke Management and Flood Control) to estimate the costs for upgrading Vietnam's coastal defence systems to achieve improved safety levels and quality of life. According to the plan, US$0.7 billion is necessary for improving sea, river and estuary dykes. Another US$5.8 billion is necessary for upgrading other defence measures such as raising houses and installing pumps. In the protection development plan, SLR was *not* taken into account. Hence the actual costs are likely to be significantly higher.

UPGRADING OF SEA AND ESTUARY DYKES. It is estimated that to improve safety levels to acceptable design standards a height increase of 1.5–2 m is required in the north, 1–1.5 m in the south and 0.3–1 m in the central provinces. A total of 2700 km of sea and estuary dykes needs to be upgraded.

UPGRADING AND DEVELOPMENT OF OTHER MEASURES. The main measures to prevent the effects of flooding are raising lands (1800 ha; US$72.5 million), raising houses (1.3 million on 128,550 ha; US$3.7 billion) and pumping (700,000 ha, unit costs ranging from

US$700/ha in the south to US$6000/ha in the north).

BEACH NOURISHMENT. The proposed plans for beach nourishment cover only 14 km of coastline in tourist areas at a total cost of US$22 million (unit cost US$1.5 million/km).

GROINS. Existing groins are rock groins in the north and timber groins in the central provinces. A small length of coastline of no more than 50 km is nominated for groin protection at a total cost of US$14.3 million.

Conclusions

People, capital value and habitats in low-lying areas of Vietnam are presently very vulnerable to flooding. The impacts of climate change will further aggravate the pressing situation. There is a danger that the focus on rapid economic expansion and industrialization will absorb development funds needed to protect and sustain the agricultural yields necessary for Vietnam to feed its own population and meet export quotas. Appropriate measures require national and international cooperation.

The findings showed the high sensitivity of Vietnam to a rise in mean sea level that could severely impact development and growth. Vietnam's vulnerability was ranked as critical and costs of full protection measures were seen to be immense. The most sensitive areas are the Mekong and Red River Delta, the Ho Chi Minh-Vung Tau area and the Hue-Da Nang area. It is expected that this study helped to improve the way decision makers and the public are informed on climate change issues.

However, this study also had its limitations. Due to an incomplete database, some models could not be used in the research and impacts of tourism were not considered in this study.

2.12.4 Experiences and lessons learned

The VA study was practical and useful. This study assessed the vulnerability of the entire coastal zone of Vietnam to the impacts of SLR due to global warming and outlined the first steps towards Integrated Coastal Zone Management (ICZM) in Vietnam. The pilot studies at three sites – Nam Ha, Thua Thien Hue and Vung Tau – were included to provide insight into the present coastal management problems.

The project was executed by a Vietnamese team working closely together with a European team comprised of Polish and Dutch experts in coastal zone management. During the study, extensive data on physical, socio-economic and institutional characteristics of the coastal zone of Vietnam were collected. Digital maps of the entire coastal zone formed the basis for GIS analyses, which determined areas of different land-use types inundated by various flood scenarios. Further analyses provided loss and risk figures for land-use types, population and capital value. Future development trends as well as institutional, organizational and legislative arrangements for coastal zone management were also reviewed and implications analysed. During the project, the Vietnamese team gained experience in working in groups within a multidisciplinary team to survey and assess comprehensively the impacts on the coastal zone and their consequences as well as apply many different measures to adapt to the envisaged impacts.

2.12.5 Follow-up research

Vietnam urgently needs to increase its coping capacity associated with SLR. The present coastal defences are inadequate to ensure the necessary safety levels for sustainable growth in the coastal zone and increasing flooding, erosion and salinity intrusion will hamper development in this zone.

ICZM can be used to focus needs and disperse ideas, data and initiatives. It is proposed that a Coastal Zone Secretariat (CZS) should be established in Vietnam, operating within the framework of a proposed Strategy and Action Plan for ICZM. Main themes of the long-term proposal are as follows:

- dissemination and exchange of information;
- harmonization of assessment and planning methodologies;
- development and dissemination of decision support and planning tools;
- training: conferences, workshops and seminars;

- coordination of bilateral and multilateral development projects.

The plans drafted by the Vietnamese government Ministry of Agriculture and Rural Development address the upgrading of Vietnam's defences to achieve improved safety levels and improved quality of life. They do not take into account the effects of SLR. The total cost of upgrading described is summarized as US$0.7 billion for sea, river and estuary dykes and US$5.8 billion for upgrading of the defence measures (including raising houses, pumping, etc.).

A joint funding agreement between the Vietnamese government and the World Food Programme (WFP) has been set up for upgrading a total 815 km of sea and estuary dykes in two programmes at a cost of US$66 million, of which US$38 million will be provided by WFP and the remainder by the Vietnamese government. Additional funding from other NGOs (Oxfam, Interchurch Organization for Development Co-operation (ICCO)) is being implemented on a smaller scale (US$2.5 million). This still only deals with upgrading of 30% of the total sea and estuary dyke system.

mentation strategy where coastal protection and management was proposed as a short-term option.

Vietnam does not have enough funds to develop new adaptation study projects. It is considered that further projects must mainly be funded from external sources, with the national budget being used as a partial contribution or for other urgent needs.

Four principal policies are as follows:

1. Greenhouse gas (GHG) mitigation.
2. Adapting to climate change.
3. Legislation and policy formulation.
4. International networking.

Some climate change policies are shown in Box 2.9.

As mentioned above, there are some methodologies and studies such as GIS analysis, that are in line with policy needs. However, they concern physical and environmental issues rather than institutional or socio-economic issues. The institutional, organizational and legislative issues should be emphasized more in future ICZM studies in Vietnam.

2.12.6 Policy implications

There were no detailed policy measures that followed the VA study, except for the imple-

2.12.7 Conclusions

Legal, institutional and organizational aspects are at present a serious threat to good

---

Box 2.9. Principal policies of the Vietnam vulnerability assessment study on the implementation of UNFCCC.

Rebuild agricultural plans for every region taking into account climate trends
Research/implement new practices
Research climate change mitigation methods, production technologies
Apply international experience and technologies as appropriate
Build new reservoirs for hydropower plants
Improve drainage of low-lying areas (more pumps)
Improve quality and protection of dykes, sea defences, estuary dykes etc.
Improve coastal protection by mangroves, beach defences etc.
Expand the hydrometeorologic station network
Improve forecasting and warnings (e.g. typhoon warnings, flood warnings, etc.)
Strengthen the ability of flood and drought committee to cope with emergencies
Proclaim decrees and guidelines for climate change response
Enforce measures to deal with violations of climate change related provisions
Proclaim law on the exploitation and protection of forests, mangroves, etc.

coastal zone management in Vietnam.[21] There is a lack of written legislation governing specific activities and developments in the coastal zone. Many organizations can contribute to a better understanding of the coastal zone but coordination and communication are weak, duplication of efforts is often encountered and no clear framework of roles and responsibilities exists beyond the hierarchy structures of the present party and government system. There is a high degree of 'vertical' decision making with a low degree of consensus at lower authority levels. The staff members are well motivated and have excellent organizational abilities. Working conditions and facilities are quite good.

Economic and financial aspects are the most serious of all problems. The present capital spent on coastal zone defences is far short of that required to meet the objectives of improved safety levels, even without SLR. The danger of flooding is rapidly becoming an obstacle to development.

The enormous funding required for upgrading of coastal defences, even without SLR, is a very heavy burden on Vietnam. It is not at all feasible to expect that even the development plan for coastal zone flood prevention will be significantly carried out. This means that unsafe dyke levels and frequent flooding problems will persist for at least several decades.

Important adaptation measures include the increase of pumping capacity and elevation of houses. Particularly in the Mekong Delta, raising houses is now the only feasible option. Increasing pumping and raising embankments to lengthen the summer crop season has been very successful but this has to be carefully managed to avoid serious upstream or downstream hydraulic and water quality problems.

The technical level of staff and facilities for ICZM studies is reasonable. Further training and experience is required to improve knowledge of coastal zone processes and coastal zone protection methods. Data availability is not a serious problem in the sense that a large amount of good coastal zone management data exists but exchange of data and readiness for distribution is limited.

There is a danger that the focus on rapid economic expansion and industrialization will absorb the development funds needed to protect and sustain the agricultural yields for Vietnam to feed its own population and meet export quotas.

In summary Vietnam's overall vulnerability level to a 1 m SLR over the next 100 years is *critical*! The VA project has been recognized and appreciated within Vietnam and constitutes a fertile contribution to further development.

Main Publications From the Research

Harris, F.R. (1994) *Vietnam Coastal Zone Vulnerability Assessment and First Steps Towards Integrated Coastal Zone Management.* Project Document (Uitvoerings document), IBW-PAN, Delft Hydraulics, The Netherlands, October 1994.

Vietnam Coastal Zone Vulnerability Assessment and First Steps Towards Integrated Coastal Zone Management (1995) *Report No. 1, Inception Report.*

Vietnam Coastal Zone Vulnerability Assessment and First Steps Towards Integrated Coastal Zone Management (1995) *Report No. 2, Data Collection.*

Vietnam Coastal Zone Vulnerability Assessment and First Steps Towards Integrated Coastal Zone Management (1995) *Report No. 3, Methodology Report.*

Vietnam Coastal Zone Vulnerability Assessment and First Steps Towards Integrated Coastal Zone Management (1995) *Report No. 4, Pilot Study – Sea Dyke Erosion in Nam Ha Province.*

Vietnam Coastal Zone Vulnerability Assessment and First Steps Towards Integrated Coastal Zone Management (1995) *Report No. 5, Pilot Study – Flooding and Lagoon Management, Thua Thien Hue Province.*

Vietnam Coastal Zone Vulnerability Assessment and First Steps Towards Integrated Coastal Zone Management

---

[21] Since the conclusion of the studies described here, much has improved. See also Subsection 4.3.1.

(1996) *Report No. 6, Pilot Study – Coastal Management and Planning, Baria-Vung Tau Province.*

2.13 Yemen

Mohamed Said El-Mashjary[22]

2.13.1 Introduction

Yemen signed the United Nations Framework Convention on Climate Change (UNFCCC) in 1992 during the Earth Summit in Rio de Janeiro. The parties participating in the Conference of the Parties were required to submit the Initial National Communication of their respective countries. In August 2003, through its cabinet the Yemeni government approved the Kyoto Protocol and the necessary documents were sent to be deposited at the United Nations' headquarters to be processed and approved.

Yemen received funds from the government of the Netherlands through the NCCSAP and from the Global Environment Facility (GEF). Climate change was a new subject in Yemen and consequently the capacities in this field are very limited. Capacity to assess topics on climate change, particularly the main focal areas of greenhouse gas (GHG) inventories, mitigation analysis, and vulnerability and adaptation assessment remains restricted to a few research institutions.

In this section we focus on the results from the vulnerability and adaptation study in the agricultural sector as this is the economic mainstay in Yemen, contributing about 20% to its GDP. Cereals, mostly millet and wheat, are grown in about 60% of the cultivated land, but a large portion of the country's import bill is reserved for wheat, grain and flour. Agricultural production in Yemen depends primarily on farm natural resources that are vulnerable to climate change. The objective of this study was to identify and evaluate possible adaptation measures that will enable the agricultural sector to cope with climate change.

2.13.2 Approach

The agriculture study team focused mainly on the effects of elevated CO_2 concentrations and higher temperatures on the economic production and physiological mechanisms of wheat and potatoes in two main realms of Yemen. The Agriculture Research and Extension Authority (AREA) in Dhamar developed the climatic database. Daily weather data were collected for assessing rainfall and temperature patterns and used for climate modelling. Detailed data were especially needed as rainfall patterns in semi-arid conditions can best be analysed using 10-day averages. A frequency analysis of the number of frost days requires daily data as the extremes are levelled out in 10-day or monthly averages. The rainfall data and number of frost days were used to predict potential yields of crops and to calculate irrigation schedules.

Four sites were chosen to represent the two main ecological zones where both wheat and potatoes are grown. The selected zones and sites were:

1. The mountain highland zone, study sites Dhamar and Ibb.
2. The desert plateau zone, study sites Mar'ib and Seyoun.

These are shown on the map of Yemen in Fig. 2.14. The studied zones represent well-pronounced variations in rainfall and temperature. The highland climate ranges from semi-arid (Dhamar) to sub-humid (Ibb); whereas, the plateau climate ranges from arid (Mar'ib) to hyper-arid (Seyoun).

The IVM/UNEP *Handbook on Methods for Climate Change Impact Assessment and Adaptation Strategies* (Feenstra et al., 1998) and the *IPCC Technical Guidelines for Assessing Climate Change Impacts and Adaptations* (Carter et al., 1994) were used as guidance in the vulnerability and adaptation assessment.

---

[22] Environment Protection Authority, PO Box 8167, Hasaba, Sana'a, Yemen. Tel./Fax: 00967 1 670238, Mobile: 00967 73761109, Email: mssmashjary@yahoo.com

Fig. 2.14. Map of Yemen.

2.13.3 Results of climate change on agriculture in Yemen

The agriculture team used semi-quantitative methods including biophysical and socio-economic assessments. A model using Microsoft EXCEL was adopted to assess relevant effects. The FAO CROPWAT (Smith, 1992) model was used to assess the impacts of three climate scenarios: the *core* scenario (Oregon State University, OSU), the *wet* scenario (Max Plank Geostrophic Ocean, ECHAM3 TR) and the *dry* scenario (United Kingdom Meteorological Office, UKH1). In addition, assessments of incremental changes were combined with observed climate data of rainfall and temperature to construct additional climate change scenarios.

In the model, wheat and potatoes yield reductions (YR) in rainfed areas and net irrigation requirements (NIR) in irrigated areas were predicted for the year 2050, under the three climate change scenarios. The results represented the combined effect of changes in both temperature and precipitation. They were compared with yield reduction and net irrigation requirement results using observed climatic data.

According to CROPWAT estimates for actual water use for potatoes in the present situation are approximately 10% more in Ma'rib than in Seyoun, a result that could be mainly associated with higher radiation load and wind speed and relatively drier air prevailing in Ma'rib. Wheat shows almost no vulnerability to climate change. This conclusion may support the idea that wheat production in the Ma'rib area will have substantial potential.

Other signs of crop production vulnerability under expected changes in temperature and precipitation could be related to the incidence of pests and the soil conditions. Increased temperatures coupled with the same or higher precipitation generally provide better conditions for the development of diseases and other pests including weeds. Physical conditions of the soil are sensitive to potential climate change. If precipitation increases and temperature either remains constant or increases, the permeability of soils in the semi-humid region (Ibb) will change. This will thus expand the area of land vulnerable to flood hazard and poor drainage.

The following sections present the data and results and give suggestions for follow-up studies.

Main results

WATER. In Yemen water is extremely scarce. Therefore groundwater consumption is very high. The annual decline of aquifer levels in most water basins averages 1–8 m. All over the country, current water quantities pumped are estimated at 138% (2.8 billion m^3) of the annual renewable quantity, which is estimated at around 2.1 billion m^3. In the mountains, extraction is as much as five times the quantity of precipitation. It is projected that at current levels of extraction the water reservoirs in this region will dry up in a period of 50 years.

Water is the limiting factor for crop production. In the irrigated areas wheat production is out of the storage risk, with no direct effect on the yield, and the water requirement is 450 mm over the growing season. No wheat production can be obtained in some years in rainfed areas, because the minimum water requirement is at least 250 mm. The highest water efficiency was 1.16 kg/m^3 with 400 mm water use. Generally, crop and irrigation patterns exacerbate the water scarcity and result in salinization of aquifers.

SOILS. In general, soils in Yemen are calcareous, with pH between 6.8 and 7.5. They have high levels of potassium, and low levels of nitrogen and phosphorus. Phosphorus was the primary limiting factor in Dhamar. Calcium carbonate may affect crops chemically and physiologically. Organic matter is lost in these soils; this means that the soils have low capacity for providing available nitrogen for crops. Accordingly, Yemen's soils need nitrogen supplements for all crops, except legumes.

There were difficulties in predicting whether climate change would affect soil nutrients; therefore it was assumed that there would be no effect. However, physical conditions will be sensitive to potential climate change. If precipitation increases, and temperature either remains constant or increases, the drainage ability of soils in the semi-humid areas (Ibb) will change. This will expand the area of land vulnerable to flood hazard and poor drainage.

CROPS. New and high yielding crop varieties were introduced in Yemen many years ago. These varieties require high levels of inputs such as water and fertilizers. They are very often resistant to diseases and pests compared with the indigenous or old races of wheat. The main disease problem is rust, particularly yellow and stem rust. Aphids are the major pests. They decrease the yield by about 10%, particularly under irrigated conditions. The use of pesticides to protect wheat production is uncommon in the study areas.

Wheat and potato yields varied from year to year and area to area. The average yields of wheat at the national level are less than 1.3 t/ha (1981–1998). The highest yields were in 1986, 1987, 1988 and 1989, whereas the lowest were in 1983, 1984, 1991 and 1997. The average potato yields fluctuated in the study areas. From 1990 to 1996 these yields increased in Ma'rib and Seyoun with an average of 15.4 t/ha. In Ibb the yields increased in the period 1985–1986 with an average of 20.5 t/ha, whereas the average was 12.5 t/ha in the period 1994–1996.

SOCIO-ECONOMIC IMPLICATIONS. Agriculture accounts for 58% of employment in Yemen. Therefore, the depletion and degradation of natural resources, particularly water and soil, have significant implications for the livelihood of a majority of the population. Wheat price subsidy increase may stimulate rainfed wheat growth.

It was estimated that women contribute about 60% to the total labour in the agricultural sector. Data also indicate that 96% of children in rural areas are engaged in agricultural work. Child labour has grown considerably over the last few years, which is considered a significant socio-economic problem.

LIMITATIONS AND DISSEMINATION. In Yemen very few studies were conducted in the field of climate change and its impacts on natural resources. The main reason for this is the lack of systematic and reliable meteorological, climate and other relevant data that are critically needed to conduct such studies. Furthermore, local experts and researchers in the field of climate change are very scarce and have difficulties participating in international activities and programmes related to global climate change.

Yemen has significant financial constraints in conducting studies, surveys and gathering scientific data. Only three studies were conducted over selected sites to assess the negative impact of climate change and the results seem to be very generalized and not very representative. Therefore, much effort is needed to complete the studies and cover more study areas to further understand the impacts and adaptation at the national level. In spite of these problems encountered during Phase 1 of the project, public awareness in climate change issues has risen and improved significantly. This was due to conducting seven awareness workshops and seminars in various cities and towns within the country.

2.13.4 Experiences, lessons learned and follow-up research

The *Initial National Communication* (EPC, 2001) in Yemen showed generalized results and findings, which were useful in fulfilling Yemen's commitment towards the UNFCCC and the Kyoto Protocol. It also helped with the establishment of a database within the Environmental Protection Authority (EPA) about Yemeni expertise in these fields and collection of a reasonable amount of data and relevant references. Furthermore, it raised the awareness of the public about the impacts and threats of climate change.

However, there is a need for conducting more research and studies to understand clearly the effects of climate change on agricultural productivity. As noted, there used to be no concern about the effects of climate change on sensitive crops, including pulses, horticultural and cash crops. It is important to understand the sensitivity of these crops to climate change, which could affect the agricultural productivity of the country, particularly in remote farmlands. It is of crucial importance to understand how the projected consequences of climate change would affect agricultural productivity.

Change in climate may lead to a changed subsidy policy and increase demands of agricultural inputs for lands, particularly pesticides and fertilizers. These resources are scarce, especially in remote rural areas. Most farmers lack funds for new investments for adapting to climate change. Adaptation policies can make sense already in the case of drought and floods because of climatic variability and extreme events. Adaptation can help to reduce damage in the short term, regardless of any longer-term changes in climate. It is of vital importance to introduce appropriate water-saving irrigation practices to contribute to the mitigation of effects of climate change on agriculture.

2.13.5 Policy implications

The Yemeni government adopted several policies that in some cases led to efficient and fruitful results. The most important policies include:

1. Subsidization of consumer prices, especially of wheat, and thus decreasing output prices, subsidization of diesel fuel and credits at low interest rates for pumping and other farm machinery, resulting in lower water prices.
2. The dominant role of the government in marketing and distribution of farm inputs.
3. Subsidies on other inputs, such as fertilizers and pesticides, fruits, vegetables and other imported grains and seeds.
4. Improvement in irrigation practices. This has led to dramatically less groundwater consumption and a significant increase in agricultural yields.

The government adopted a Gender Food Security and Agricultural Policy. Policies also address practical and strategic gender needs in training, research, agricultural extension, credits, micro-enterprise and marketing. Recently regulatory measures, such as national acts, have been issued in Yemen for the following aspects:

1. Pesticide use (the State Law on plant pesticides was issued in 1999);

2. Water use (the Water Law was issued in 2002);
3. Seeds and agricultural fertilizers (Ministerial regulations and internal circulars):
- to register kinds of certified seeds including seeds and seedlings;
- to regulate seed production and marketing;
- to protect ownership of invention patents;
- to regulate agricultural fertilizer use and protect humans, animals and the environment from their direct and indirect adverse effects.

The methodologies and studies used by the study team were in line with the policy needs, however these methodologies could be developed further or additional methodologies could be introduced in follow-up studies. The aims of these policies are to introduce subsidies or taxes to encourage the growth of crops that are more resistant to the (future) climate conditions. Many farming technologies, such as efficient irrigation systems, provide opportunities to reduce dependence on natural factors such as rainfall and runoff. Also, the effects of market liberalization must be studied further in light of climate change.

Main Publications From the Research

Environmental Protection Council (EPC) (1999) *Report on Climate Change Country Study in Yemen and Enabling Yemen to prepare its First National Communication in Response to its Commitments to the UNFCCC.* EPC, Sana'a, Yemen.

Environmental Protection Council (EPC) (2000) *An Assessment of Climate Change on Agriculture in the Republic of Yemen.* Final report submitted by the Agriculture Team to EPC/UNFCCC. EPC, Sana'a, Yemen.

Environmental Protection Council (EPC) (2001) *Initial National Communication under the United Nations Framework Convention on Climate Change.* EPC, Sana'a, Yemen, 72 pp.

2.14 Zimbabwe

Washington Zhakata[23], Norbert Nziramasanga[24] and Margaret Sangarwe[25]

2.14.1 Introduction

Zimbabwe is the second largest energy user in Southern Africa following South Africa. The regional energy sector is dominated by coal with hydroelectricity being available mostly in countries north of Zimbabwe. The Zambezi River is the major source of hydroelectricity at present in Zimbabwe but the Inga Falls on the Zaire River offer the greatest potential for hydroelectricity in Africa. The current domination of coal is expected to continue due to its low costs. Since the use of carbon-intensive fuels is dominant there is scope for initiating activities to optimize energy use in the country.

Zimbabwe is also prone to frequent droughts and in recent years flooding has affected communities in the south-eastern parts of the country. This in a way reinforces the need for the country to participate in activities that may mitigate the impacts of the changing climate. Therefore, a study on climate change impacts and adaptation in the agricultural sector and mitigation opportunities in the energy sector has been conducted. In this section we will focus on the greenhouse gas (GHG) emissions mitigation assessment activities.

We will emphasize cogeneration in Zimbabwean industries. In brief this NCCSAP study sought to identify win-win opportunities in industrial energy efficiency improvement where companies, despite knowledge of the opportunities, have not been able to implement such opportunities. Hence, the study identified the implementation barriers and tried to make recommendations for barrier removal.

---

[23] Climate Change Office, Ministry of Environment and Tourism, Private Bag 7753, Causeway, Harare, Zimbabwe. Tel.: +263 4 757881-5, Fax: +263 4 757876, Email: climate@ecoweb.co.zw
[24] Southern Centre for Energy and Environment.
[25] Ministry of Environment and Tourism.

2.14.2 Approach

The activities in Zimbabwe were centred on the identification of opportunities for climate change mitigation through emission reduction. The focus was on industrial energy use with a limited exercise to consider the potential for utility level efficiency improvement. The project concept was to identify opportunities where energy efficiency improvement would yield a win-win solution but where the company had not implemented the measure, i.e. the opportunity would save both money and the environment. The main objective was to identify the barriers preventing implementation of the measure and how these barriers could be removed. As a result it was evaluated to be more strategic to identify options that had been analysed earlier since these would be interpreted for the companies and might show ways for them to overcome the barriers of implementation.

Zimbabwe has had a significant number of studies in the context of energy efficiency improvement and climate change. A study carried out under the support of the Canadian International Development Agency from 1989 to 2001 looked at industrial energy efficiency and how capacity could be built within local institutions and individuals to implement efficiency improvement measures in the Southern African Development Conference (SADC) industries. Zimbabwe was part of this study and several training courses were run which resulted in the identification of options for energy efficiency improvement. Other studies carried out in preparation for the First National Communication to the United Nations Framework Convention on Climate Change (UNFCCC) also yielded several options for energy efficiency improvement. In more recent years the United Nations Industrial Development Organization (UNIDO) sponsored activities to 'learn by doing' in building capacity for implementing Clean Development Mechanism (CDM) projects (see also Subsection 4.2.2). These studies also sought to identify barriers to efficiency improvement or implementation of CDM projects by conducting workshops and raising awareness on institutional roles in implementing CDM. All these activities linked very well with NCCSAP by providing background information and by pointing consultants to cases with the greatest potential.

Several consultants carried out this study. Each consultant was responsible for a case study, which in most cases was a single industrial plant. Some consultants handled more than one industrial plant but with each plant representing a case study. In total nine case studies were carried out. The consultant visited the plant and compiled information on the processes used and the energy use patterns. Where possible the consultant discussed the opportunities for energy efficiency with the plant management or technical people. The objective of the discussions was to identify the major barriers that were preventing the implementation of win-win opportunities that were otherwise well understood.

After data collection the consultant built an option for energy efficiency improvement and tried to quantify the potential for GHG emission reduction and the other local and global environmental benefits. Where there was not sufficient technological data the consultant got assistance from the technical support team that included the Southern Centre for Energy and Environment, the Institute for Environmental Studies (IVM) and the Energy Research Centre (ECN).

Two workshops were held where the consultants presented their findings and got feedback from the participants. This interaction allowed the stakeholders to comment on the work and also to guide the consultants in data collection and analysis. In the Bindura Smelter and Refinery case the stakeholders developed a priority list of the opportunities that were studied. We will focus on this case in the following subsection.

2.14.3 Results of the Bindura Smelter and Refinery study

Zimbabwe has had experience with steam/electricity cogeneration in the sugar industry. All other industries have not previously considered cogeneration as an option that they could successfully pursue. The Zimbabwe iron and steel company, ZISCO, operates a power station fired on blast furnace gas. This is similar to cogeneration but is more of a power plant taking advantage of waste energy from the blast furnace without much need for coordination between fuel production and use since the plant can be idle when the blast furnace is operational. In any case the blast furnace gas can be used for other heating operations in the mill. The identification of the cogeneration opportunities at the Bindura Smelter and Refinery helped in highlighting the potential for new cogeneration opportunities that seek to address national problems of coal supply and security of electricity supply. The plant has an ideal situation because the existing boilers need to be replaced and the cost of coal transport has risen because of a poor transport network. In addition, grid electricity supplies are intermittent due to limited internal generation capital.

The Bindura Smelter and Refinery processes nickel ore from the adjacent nickel mine. The bulk of the energy used in the smelter is for heating ladles and for producing steam used for maintaining the high temperature in the electrolysis plant. The company is keen on energy efficiency improvement since the energy input presents one of the major production costs. In addition the company exports all its output and the international nickel prices are determined outside the influence of the plant. Efficiency improvement is therefore a key option for increasing margins.

The options for efficiency improvement include upgrading the smelter technology to include energy recovery from the sludge or to employ high technology smelters where the heating effect would be more accurate, e.g. with plasma based processes. At a simpler level the boiler efficiencies could be improved by replacing the existing shell type boilers with more modern units with higher design efficiencies and flue gas heat recovery. An option that presented itself earlier, but was only analysed later, was the production of electricity from the steam produced by the boilers before sending it to the electrolysis plant. The electrolysis plant accepts low-pressure steam at a low temperature. The boilers are capable of high-pressure steam but are operated at 10 bars, which suits the electrolysis plant.

The low steam pressure reduces the energy efficiency for the process. Hence the smelter uses more coal than is necessary and at the same time it is drawing electricity from the grid for driving the conveyors, electrolysers and other process equipment. The design efficiency of the existing boilers is about 74% while new boilers could have an efficiency as high as 85%. The existing boilers were installed in 1969 and 1974 and are now due for replacement. It is in this context that installation of a cogeneration unit is being considered. The boilers would be replaced by higher-pressure units of preferably 40 bars. A backpressure turbine would be installed and the exhaust steam from the turbine would be used to heat the electrolysis plant. A quick calculation showed that the turbine would produce about 3 MW of electricity. This would be about 20% of the total plant load, which is 15 MW. The use of high-pressure steam results in the production of additional energy from the same quantity of coal. This would also reduce the need to transport coal. Road transport has increased dramatically in the last few years at the cost of rail transport for moving coal from the coal mine to the plant.

The proposed option (see Fig. 2.15) was well received by the plant management who thought it offered a realistic possibility for them. What appeared to be the major barrier was the lack of experience with cogeneration of this type.

The issue of data and information on the proposed cogeneration plant presented a problem for the consultant. The cost of equipment was not readily available and manufacturers tend to compute costs on a case-by-case basis, which would require more detailed design. The project did not provide for this, as the objective of the project was to get indicative figures. In an attempt to acquire cost figures a search for a similar plant was made on the Internet. The findings were mostly for individual pieces of equipment, such as turbines or alternators, and not for a complete plant. The costs of installing equipment vary significantly from site to site. Hence it would have been better for the project to apply cost figures from a complete project as opposed to adding individual unit prices. Eventually generic figures were used. These were based on the known average cost for a thermal fired plant.

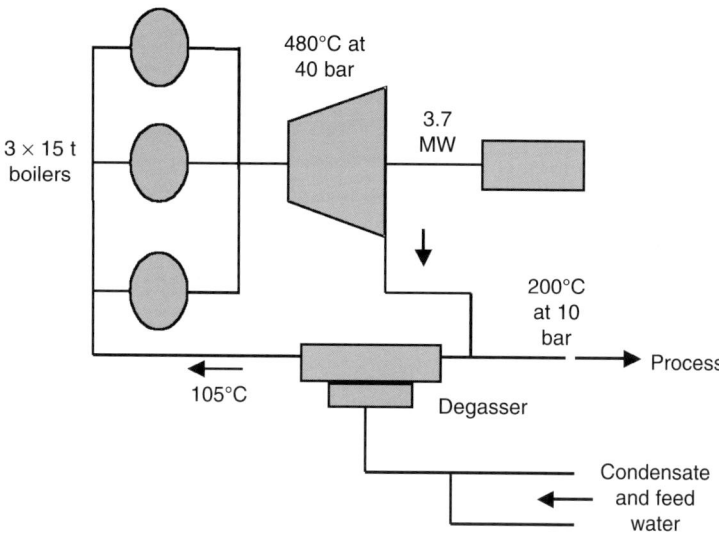

Fig. 2.15. Proposed cogeneration layout.

Zimbabwe has a high potential for the installation and use of cleaner technologies with win-win benefits. Despite several studies having been carried out there are major impediments to the immediate implementation of these technologies. The barriers are elaborated below.

Financial barriers

The costs of new technologies are not necessarily the reason for slow adoption of cleaner technologies. Companies seem to be focused on other urgent issues and other, more urgent technologies have been implemented.

Due to the current macroeconomic situation in Zimbabwe, industry is facing very high inflation and interest rates that make it difficult for companies to secure new loans for investments in modern, more efficient, technologies. In addition, the devaluation of the currency rate also makes imports more expensive. These factors lead to a high level of uncertainty on future prices and costs and investors therefore tend to delay their investments.

Awareness of cleaner technologies

In cases where companies have installed new technologies they have not considered additional components that could have achieved cleaner performance with lower cost. This includes the turbine installed at Hypo Valley (where higher-pressure steam would have increased the volume of power produced) and the new pumps installed at Bulawayo pumping stations (where upgraded controls could have allowed for optimum pump sizing as opposed to the traditional 10% over sizing). This in a way points to a lack of up-to-date information on technology improvements.

Limited technology assessment skills

Motivation to upgrade technology is often a result of existing skills to assess the benefits and present a convincing story to the decision makers. Climate change is not so high on the corporate agenda, hence technical staff in industry do not spend time looking for climate friendly opportunities. As a result, skills are not developed and technology assessment for environmental protection is not used as a criterion for investment.

2.14.4 Experiences and lessons learned

The study was structured to rely on previous work for identification and analysis of energy efficiency improvement options.

In the absence of previous work the study was limited in terms of identification of barriers to implementation of the options since the potential investor would not have had knowledge of the option.

On the other hand, there was room to discuss with the company management in an attempt to identify what the barriers could be, provided the management was in a position to respond quickly to a new idea. Relying on existing reports required that the technical staff in the company be familiar with the studies. Where this was the case it was easy to identify the barriers. In other cases the staff did not offer suggestions on what the barriers could be. Sable Chemicals is one case where staff were familiar with the options. It was therefore fairly easy to hold a discussion on the potential barriers with the technical managers. This also applied to Valley Estates where the company had installed a 20 MW turbine and was in the process of negotiating a power purchase agreement with the utility. In the case of Bindura Smelting and Refinery and the pumping stations in Bulawayo it was much more difficult to discuss barriers because the opportunities were just being identified.

Technology information was limited for all the case studies. Data were generally available in the form of nameplate data for machines but more detailed information on operating parameters and alternative technologies was generally not available. The key technical contacts in the companies were also limited in their knowledge of this information. The project requires a multidisciplinary team where, apart from appreciation of energy efficiency improvement, the team would need to be able to make preliminary designs to enable determination of capital requirements and other technical data. This missing data restricted the ability of teams in determining the emission levels and the potential benefits of efficiency improvement. Therefore it was concluded that the findings could be used as a basis for more precise analysis in future.

The problem is that boiler costs needed to be split into the cost for a normal boiler like the one already installed and the cost of a higher-pressure boiler such as the one proposed. This would yield the incremental cost of the improvement. This calculation could not be done, however, because boiler manufacturers in the region could not provide this information since they do not manufacture boilers with pressure above 20 bars as a standard product. Given the interest shown by the company management it would be important for them to seek the services of a cogeneration expert so they can get a more detailed analysis of the option with more accurate cost and engineering design figures.

The workshops and other discussion forums that were held showed that there is general interest in energy efficiency improvement in industry. Even though there was no immediate knowledge of policy options to encourage the implementation of identified measures there was a general agreement that something has to be done at a high level to encourage implementation of the measures.

2.14.5 Follow-up research

While there are some aspects of climate change driven policy that are unique, it has been identified that many of the activities of a successful climate change technology transfer programme provide benefits towards a broad range of development objectives, and the lessons learned through the needs assessment process can be applied to a range of development challenges where technology transfer and international cooperation can contribute.

The NCCSAP mitigation studies have just been completed. Follow-up research areas have been identified in order to implement the recommendations that were identified in the main study. It has been identified that Zimbabwe has a high potential for the installation and use of cleaner technologies with win-win benefits. Despite several studies having been carried out there are major impediments to the immediate implementation of these technologies. Follow-up research needs to be focused on finding means of addressing the issue of slow adoption of cleaner technologies. Companies seem to be focused on other pressing issues so that there is an absence of motivation to concentrate on optimization of activities including installation of technology for cleaner production reasons. One area that needs researching is how indigenous capacities in the country can be enhanced in order to develop and transfer technologies.

From the studies, it also emerged that there is a need to explore the ways in which companies can consider installing additional components that could achieve cleaner performance with lower cost.

One of the most important future research areas as a follow up to the NCCSAP is on the issue of assessing the levels of motivation to upgrade technology. This will entail the establishment of collaborative partnerships between key stakeholders with the common purpose of enhancing technology transfer. A way has to be found for the government to be involved at all stages. Enhancement of the capacity of resident skills to assess the benefits and solicit commitment of decision makers will be paramount. It was observed in the study that climate change is not so high on the corporate agenda, hence technical staff in industry do not spend time looking for climate friendly opportunities. As a result skills are not developed and technology assessment for environment protection is not used as a criterion for investment. There is a need to design an integrated programme for barrier removal that would see accelerated upgrading of technology in industry.

2.14.6 Policy implications

The climate change issue is a new concept in Zimbabwe in terms of both its science and its policy implications. The understanding of this subject – its scientific basis, institutions and relevance to Zimbabwe's economy – are mainly restricted to a few institutions and individuals working on the subject. It is not possible, therefore, that Zimbabwe would, at this stage, have a stated or fully considered national perspective on policies and measures to respond to climate change. However, climate change activities in Zimbabwe have been growing since its participation at the Earth Summit in Rio de Janeiro in 1992.

It is worth noting that recommendations to incorporate climate change policies into Zimbabwe development plans have been captured in the new act on the environment 'Environment Act of Zimbabwe' of 2003. What is significant at this point is that Zimbabwe has included climate change in its national legislation. The NCCSAP project results were well circulated among various stakeholders and have played a significant role in the formulation of climate change policy in Zimbabwe.

2.14.7 Conclusions

With the experience gathered from these studies, Zimbabwe is now in a better position to conduct more comprehensive assessments of GHG emission inventories, mitigation analyses and vulnerability and adaptation studies.

There are viable opportunities for climate change mitigation in Zimbabwean industry. These opportunities come with major barriers to implementation, which require joint effort between public and private sector decision makers to overcome. Even though studies have been carried out that analyse these barriers there is a need to solicit commitment from all parties for the successful implementation of these opportunities.

Several barriers appear to limit the adoption of win-win technologies in Zimbabwean industry, such as lack of cost information, lack of interest of plant managers and hyperinflation. As a result, policy change, awareness building and energy pricing are taken in isolation and cannot achieve the desired results of cleaner technology. There is a need to implement an integrated programme for barrier removal that would see accelerated upgrading of technology in industry. The programme could start by pursuing those options that are more promising in terms of financial and economic benefits as well as in terms of wider market application.

3
Cross-country Syntheses

3.1 Introduction

In this chapter the technical consultants compare the study results in all countries sector by sector. They summarize the main results and explain the similarities and differences between the countries. This chapter includes the following sections: Emission Inventories (3.2), Mitigation Assessment of the Energy Sector (3.3), Adaptation and Water Resources (3.4), Adaptation in Coastal Zones (3.5), Adaptation and Land Use (3.6) and National Communications (3.7).

3.2 Emission Inventories

Michiel van Drunen[1]

3.2.1 Overview

An overview of the main results from the emission inventories, linked to some general information about the countries, is presented in Table 3.1.

Table 3.1. Comparison of various key characteristics of the three countries studied in 1994 (1 Gg equals 1000 t).

	GDP/cap ('94 US$)	Population (million)	Net GHG emissions per capita (t CO_2-eq)	Net total GHG emission (Gg CO_2-eq)	Largest GHG contribution	2nd largest GHG contribution
Bolivia	870[a]	7.5[b]	8[b]	61,163	Land use change (65%)	Agriculture (18%)
Kazakhstan	2,442[c]	16.2[c]	15	212,613	Energy (89%)	Agriculture (8%)
Suriname	916[d]	0.40[d]	14	5,626	Land use change (52%)	Energy (36%)

[a]Data for 1995 from the World Bank Group; taken from www.worldbank.org, Nov. 2000.
[b]1994 population estimated at 7.5 million, from the CIA year 2000 estimates of population (8.2 million) and population growth rate (1.83%).
[c]KazNIIMOSK (1998).
[d]Suriname Central Bank and Suriname Central Bureau of Statistics. As found on www.suriname.org, Nov. 2000.

---

[1] See Note 1, Chapter 1.

The per capita emissions are not very low compared to Annex I countries. For example, in 1994 the per capita greenhouse gas (GHG) emissions in the EU-15, The Netherlands and the USA were 10, 15 and 24 t CO_2 equivalents, respectively. However in these countries most of the emissions result from the energy and industry sectors.

In Table 3.2 the GHG emissions by sector in the three countries investigated are summarized. Kazakhstan, as the most industrialized country, has by far the highest CO_2 emissions, as can be seen from the contributions from the energy sector. Bolivia and to a lesser extent Suriname emit significant amounts of GHGs from land use change, mainly resulting from deforestation. In Kazakhstan, there was a net uptake of CO_2 by plants or trees in managed areas (6672 Gg).

In all countries there are significant contributions of CH_4 emissions to the total amount of GHG emissions: 22%, 19% and 15% for Bolivia, Kazakhstan and Suriname, respectively.

3.2.2 Bolivia

In 1994 the total GHG emissions in Bolivia were estimated to be 61,163 Gg CO_2 equivalents. CO_2 emissions related to deforestation accounted for by far the largest GHG contribution. CO_2 from deforestation made up 65% of weighted GHG emissions and 83% of total CO_2 emissions. Compared to this all other sources were relatively small. CH_4 emissions, mainly from agricultural sources, made up 22% of GHG emissions and the contribution of N_2O was marginal (1.3%).

When looking at the distribution over the sectors responsible for the GHG emissions, the figures are also remarkable: energy and agriculture played a relatively small role in the total GHG production (16% and 18%, respectively) and deforestation related emissions accounted for 65% of GHG emissions. However the uncertainty of the latter contribution is estimated to be 35%. The industry and waste sectors both contributed less than 1%.

3.2.3 Kazakhstan

An important factor to take into account when assessing the emissions of GHGs for Kazakhstan is the economic decline that took place over the period studied. From 1990 to 1994 Kazakhstan's GDP decreased by 40%. This has resulted in an overall decrease of GHG emissions of 33%.

The largest contributions to GHG emissions by far (89% in 1994) were emissions,

Table 3.2. Comparison of the GHG emissions by sector of the three countries studied (1994 figures in Gg CO_2 equivalents).

	GHG	Bolivia[a]	Kazakhstan[b]	Suriname[c]
Energy	CO_2	7,646	178,252	1,996
	CH_4	1,870	17,735	3
	N_2O	62	40	3
Industry	CO_2	394	1,014	48
Agriculture	CH_4	10,275	17,388	629
	N_2O	536		6
Land use change	CO_2	38,617	−6,627	2,718
	CH_4	1,148		197
	N_2O	118		19
Waste	CH_4	429	4,811	7
	N_2O	68		
Total	CO_2 eq	61,163	212,613	5,626

[a]Source: Ministry of Sustainable Development and Planning (2000).
[b]Source: KazNIIMOSK (1998).
[c]Source: Becker et al. (1999).

particularly CO_2, from the energy sector. The energy production sector relied heavily on coal (88% in 1994), which typically has a high CO_2 output per unit energy. The second largest contribution to GHG emissions (in CO_2 equivalents) was from agriculture, mainly as a result of CH_4 and N_2O emissions.

3.2.4 Suriname

An important aspect of the emissions characteristics of Suriname is the use of energy (both fossil fuel and hydroelectricity) by the bauxite industry, which consumes 73% of all the energy in the country. A larger source of CO_2, however, is deforestation, which accounts for 52% of the emissions. The CH_4 emissions are 75% attributable to agriculture (rice paddies and livestock). Contributions of other GHGs are negligible (<1%).

In the emission inventory (Becker et al., 1999), no attempt was made to estimate uncertainties. The uncertainty of the total emissions probably amounts to at least the percentage indicated by Bolivia for the contribution of land use change (35%).

3.2.5 Discussion and recommendations

Data availability

All the countries that conducted an emission inventory had used the Intergovernmental Panel on Climate Change (IPCC) methodology as described in Subsection 1.3.2. Although this was considered a good methodology in general, most countries had difficulties with some aspects of the IPCC methodology.

First of all, activity data were not always available. This was especially the case for land use change and forestry, and waste and biomass consumption. The privatization of government institutions was given as one reason for the difficulty in gathering data.

A second problem that was encountered by the countries was a poor fit of the bottom up and top down methodologies. The main reason for this is probably the aforementioned lack of data, but country-specific circumstances can also increase the gap between the two approaches. In the case of Suriname for example, fuel smuggling creates a discrepancy in the data, but it is very difficult to deal with this in a systematic way. Furthermore, the IPCC guidelines do not cover this particular case.

Emission factors

On the issue of emission factors, it was generally acknowledged that region-specific emission factors are needed, especially for livestock, coal production/consumption and land use. To generate these emission factors, a few suggestions were made to improve access to relevant data during the Amsterdam workshop (Dorland et al., 2001). One suggestion was that groups of countries in a region that have similar problems conduct research together. Another suggestion was to improve access to relevant studies for region-specific factors by giving more attention to networking among countries, for example by making use of the Internet.

In the meantime the IPCC has set up an emission factor database and an electronic discussion group at its Japanese website (http://www.ipcc-nggip.iges.or.jp/).

Dealing with uncertainty

There were some comments on dealing with uncertainty in emission inventories. The IPCC guidelines for dealing with uncertainty were found to be difficult to adhere to. In Suriname they did not even make an attempt to make an uncertainty analysis, because they did not know how to do this based on their data and experience.

To improve the general quality and usefulness of the emission inventories, the researchers would like to be more certain that the most important sinks and sources of GHGs are indeed those that the emission inventory identifies, and that the magnitude of these sinks and sources is accurately known (Dorland et al., 2001). In Bolivia and Suriname, the most important source, land use change and forestry, has the highest degree of uncertainty. Therefore, more research is needed to increase the certainty of the figures for deforestation and reforestation. Because of the large differences between and within countries, research needs to be directed to the generation of region-specific emission and growth factors.

3.3 Mitigation Assessment of the Energy Sector

Nico van der Linden[2] and Jan-Willem Martens[3]

3.3.1 Introduction

One of the objectives of a mitigation assessment is to identify the potential of environmentally sound technologies that are required for reducing GHG emission targets in non-Annex I countries. Environmentally sound technologies can be defined as technologies that are:

- more sustainable, less polluting and more efficient than substitutes;
- compatible with national socio-economic, cultural and environmental priorities.

To achieve a sustainable application of environmentally sound technologies in developing countries, these countries require assistance with developing human capacity (knowledge, techniques and management skills), developing appropriate institutions and networks, and with acquiring and adapting specific hardware. Given the importance of taking into account the local sustainable development priorities in the technology transfer processes, it is clear that the sustainable transfer of mitigation technologies is country specific, sector specific, technology specific and user specific.

3.3.2 Analysis

In the framework of the Netherlands Climate Change Studies Assistance Programme (NCCSAP) Phase I study, assistance was provided to the national mitigation teams of Bolivia, Yemen, Zimbabwe and Mongolia (see Table 3.3 for their main characteristics). The type of assistance provided differed per country. For Bolivia and Yemen the assistance involved training on the LEAP model (see Subsection 1.4.2) and the development of a LEAP version that could be used to evaluate the identified GHG emission reduction options. The type of assistance provided in Bolivia and Yemen consisted of:

- updates of the 1990 inventory of GHG emissions;
- identification and evaluation of options to reduce GHG emissions;
- emission inventories based on the specifications of the United Nations Framework Convention on Climate Change (UNFCCC). The method applied was taken from the IPCC.

With regard to the identification and evaluation of options to reduce GHG emissions, the steps taken included:

- design of a LEAP version for Bolivia/Yemen;
- design of a baseline scenario;
- evaluation of various reduction options.

In Zimbabwe and Mongolia the analysis of mitigation options had already been completed and the activities focused on barrier analysis and policy formulation. The activities in Zimbabwe and Mongolia included:

- identifying financial and institutional barriers to the implementation of promising GHG emission reduction options;
- formulating policies and implementation strategies for GHG mitigation;
- developing promising mitigation projects.

The following observations can be made from Table 3.4:

- Total identified GHG emission reduction potentials in the energy sector in Bolivia and Yemen amount to approximately 2.8 Mt and 7 Mt, respectively. A substantial share of this amount in both countries can be achieved with negative costs (no-regret options). Improvement in energy efficiency and a fuel switch (in particular towards

---

[2] See Note 2, Chapter 1.
[3] See Note 3, Chapter 1.

Table 3.3. Main characteristics of the activities in Bolivia, Yemen, Zimbabwe and Mongolia.

	Bolivia	Yemen	Zimbabwe	Mongolia
Implementing organization	Programa Nacional de Cambios Climaticos (PNCC), Ministry of Sustainable Development and Environment. PNCC had subcontracted the Chemical Institute of the University of San Andres to conduct the emission inventory	The Environmental Protection Council (EPC). The EPC has established working groups responsible for conducting the mitigation study and the emission inventory. The mitigation group consisted of scientists from the University of Sana'a	Climate Change Office of the Ministry of Mines, Environment and Tourism. The Southern Centre for Energy and Environment in Zimbabwe was subcontracted to work on the mitigation analysis	The Institute of Meteorology and Hydrology of the Ministry of Nature and Environment. A steering committee was formed consisting of representatives of other Ministries and key stakeholders outside the government
Methodology	GHG inventory for 1994 up to 2030 Design of LEAP model for Bolivia Development of reference scenario Identification and evaluation of GHG emission reduction options Construction of aggregated cost abatement curve	GHG inventory for year 1990 Design of LEAP model for Yemen Development of reference scenario Identification and evaluation of GHG emission reduction options Construction of aggregated cost abatement curve	Identify institutional and financial barriers To draw up the plan of operations To train the Zimbabwean experts on the technical, financial and institutional aspects To comment on interim and final reports	Prepare a list of mitigation options Identify stakeholders Prepare shortlist of mitigation options jointly with the stakeholders involved Conduct pre-feasibility studies for shortlist Identify potential projects and funding possibilities Prepare implementation strategy Disseminate the information through workshops
Implementation	Duration 18 months Two missions and back-up support Delays due to capacity constraints with implementing organization Limited expertise in terms of number of staff	Duration 18 months Two missions and back-up support Delays due to political circumstances Limited expertise in terms of knowledge	Duration 25 months Three missions and back-up support Delays due to political circumstances Sufficient expertise in terms of number of staff and knowledge	Duration 22 months Three missions and back-up support Delays due to local experts being unavailable Limited expertise in terms of number of staff Considerable overlap with other donor programmes

Table 3.4. Main results of the mitigation analysis in Bolivia, Yemen, Zimbabwe and Mongolia.

Bolivia: Reduction options

Re-division of the expansion of the power sector (–100 US$/t CO_2)
Increase in the use of gas in transport sector (–25 US$/t CO_2)
Efficient lighting in the commercial sector (–20 US$/t CO_2)
Efficient commercial use of biomass (–15 US$/t CO_2)
Efficient use of biomass for cooking (–10 US$/t CO_2)
Reduction of gas burning associated with the exploitation of oil (–10 US$/t CO_2)
Energy conservation in industry (–5 US$/t CO_2)
Efficient lighting in the residential sector (15 US$/t CO_2)
Efficient cooling in the residential sector (20 US$/t CO_2)
Reduction of electricity consumption in the commercial sector (25 US$/t CO_2)
Increase of the use of solar energy (60 US$/t CO_2)
Increase of the use of gas in the residential sector (175 US$/t CO_2)

Yemen: Reduction options

Efficient stoves in the rural areas (–17 US$/t CO_2)
Efficient irrigation (–16 US$/t CO_2)
Shift to gas turbines in the power sector (–15 US$/t CO_2)
Efficiency improvements in commercial sector (–7 US$/t CO_2)
Efficient lighting in rural areas (3 US$/t CO_2)
Efficient lighting in urban areas (50 US$/t CO_2)
Efficiency improvements in transport sector (60 US$/t CO_2)

Zimbabwe: Barriers identified

Lack of financial resources – high interest rate and lack of foreign currency
Lack of government backing on some projects
Lack of sufficient data to assess feasibility and viability of possible projects
Lack of appropriate skills and know-how by operational staff
Absence of competition and hence no drive to invest in modern technologies
Existence of subsidized energy prices
Absence of a legal framework to guide negotiations of a power purchase agreement
Inability to correctly assess the benefits from the CDM

Mongolia: Contribution to the National Action Plan on Climate Change

---

A detailed pre-feasibility analysis for eight promising mitigation options: (i) application of efficient mining technology (selective mining, dewatering system, coal handling plant); (ii) coal briquetting; (iii) accelerated introduction of PV solar system; (iv) introduction of electric boilers for small, medium-sized heat boilers; (v) good housekeeping measures in industry; (vi) lighting efficiency improvements in built environment; (vii) introduction of new, more efficient small, medium-sized boilers; and (viii) introduction of steam-saving technology in industry
Gain insight into the barriers that prevent the implementation of identified reduction options
Identification of concrete projects to reduce emissions and possibilities for financing these projects

---

natural gas) appeared to be the most promising options to reduce GHG emissions.

- The main barriers identified in Zimbabwe include weak institutional framework, immature markets, limited awareness on technology information and the current political/economic situation. Through detailed energy audits carried out by local consultants at eight energy intensive companies in Zimbabwe, a number of options have been identified and evaluated to reduce these barriers. However, implementation of these options appeared difficult due to the current political situation in Zimbabwe.
- In Mongolia, the NCCSAP has supported the development of the Climate Change

Action Plan. This plan includes actions to strengthen the national capacity for climate change issues and provides strategies for the implementation of GHG reduction options for the power sector, industry and the built environment.

3.3.3 Follow-up activities

As a result of the assistance provided in the framework of the NCSSAP programme, several follow-up activities have been identified and elaborated, which are summarized below.

One of the promising options identified in Bolivia is a substitution of diesel and petrol in the transport sector by compressed natural gas (CNG) or liquefied natural gas (LNG). In Bolivia the vehicle fleet is mainly concentrated in the big cities, causing traffic congestion and local air pollution problems. The energy efficiency of these vehicles is generally very low compared to Western standards. The LEAP model calculations have shown that a conversion of vehicles powered by petrol and diesel to CNG or liquefied propane gas (LPG) is cost effective (pay back periods of 2–3 years) and also results in a substantial reduction of CO_2 emissions.

Based on the results of the mitigation study conducted in Bolivia, a new project has been formulated which aimed at a detailed assessment of the financial, technical and social feasibility of converting private and public vehicles in La Paz and El Alto to CNG and LPG. A project team has been established consisting of the Energy Research Centre (ECN), AGAS (national gas distributor in Bolivia responsible for gas distribution and gas sales), Vialle (Dutch manufacturer of LPG equipment) and the PNCC. A project proposal has been submitted to the Netherlands PPP-JI programme. The study began in 1999 and was completed in 2001.

In Yemen a project idea has been discussed with the Environmental Protection Council (EPC) involving the installation of 1000 solar home systems in the rural area of Yemen (implementation of the option efficient lighting in the rural areas). Two or three locations have been identified that are not connected to the national grid and will not be in the foreseeable future. The intention is to establish a consortium consisting of a photovoltaic (PV) manufacturer from the Netherlands, a local PV dealer, ECN and EPC. However, the required funding for this initiative has not yet been secured.

In Mongolia a draft proposal has been prepared for the establishment of the 'Sustainable Energy Centre Mongolia'. A proposal was made to set up the Centre based on an arrangement between Mongolia and the Netherlands. This was seen as very promising, as Mongolia could benefit from the Netherlands' extensive experience with energy efficiency and renewable energy policies and practices. The Sustainable Energy Centre would operate as an intermediate body between the public and private sector with the following primary goals:

- to be a centre of expertise on energy efficiency and renewable energy;
- to broker and negotiate bringing public and private stakeholders together;
- to advise the government on policy;
- to promote awareness;
- to act as a project initiator, facilitating national and international actors to set up energy efficiency, renewable energy and Clean Development Mechanism (CDM) projects.

The proposal was submitted to the Dutch embassy in Beijing and has been incorporated into the bilateral aid programme for Mongolia.

In Zimbabwe, a potential CDM project was developed together with local counterparts. The project aimed to build a 3.5 MW thermal power plant using wood residues from the nearby sawmill, which produces some 84,000 m^3 of wood residue that is underutilized at present (some wood is used for steam generation). This will ensure the disposal of the waste in an environmentally friendly manner and provide a renewable energy resource for the benefit of all stakeholders. The electricity produced by the power plant would be fed into the grid and the steam produced could be used in the production process of the sawmill. Preliminary calculations of the economics of this proposed project indicated that it could be a financially viable undertaking if brought under the CDM scheme. A project proposal for a feasibility study containing a detailed description of the project has been drawn up. However, the political situation in Zimbabwe appeared to be a major problem in getting the project funded.

3.4 Adaptation and Water Resources

Arjan van der Weck[4] and Hans Leenen[5]

3.4.1 Introduction

Vulnerability assessments and the development of adaptation strategies for water resources management have been executed in Bolivia, Ecuador, Ghana, Mali, Suriname and Yemen.

This section describes the results of the vulnerability and adaptation studies, the common risks and the identified adaptation strategies associated with water resources management in the above listed countries. The major risks are flooding and erosion due to higher river discharges on the one hand and increased water scarcity due to a decrease in average rainfall on the other hand.

Effects of climate change are very much geographically dependent and may thus differ quite widely. Flooding and resulting erosion effects can largely be combated by technical interventions, whereas combating water scarcity problems generally requires a more institutional approach. The latter is related to cooperation between users and water demand management mechanisms. The conclusion, therefore, is that in addition to technical issues, institutional issues need to be addressed in all the investigated countries that aim to prevent water spillage and achieve a higher level of cooperation between stakeholders. The concept that needs to be addressed is the recognition of water as an economic good.

Climate change will lead to more precipitation but also to more evaporation. Precipitation will probably increase in some areas and decline in others. Making regional predictions is further complicated by the extreme complexity of the hydrological cycle: a change in precipitation may affect surface wetness, reflectivity and vegetation, which then affect evapotranspiration and cloud formation, which in turn affect precipitation. Clear answers about the risks of climate change on water resources were not available at the time the studies described here were carried out.[6]

From the research in the various countries it follows that the identified risks depend on regional geography and the climate scenario considered (social, economic and cultural characteristics). The major risks may be different for the different countries; however, they can be attributed to one of two categories:

1. Increased risk of flooding and erosion.
2. Increased water scarcity.

Typically, the expected problems in Bolivia, Ecuador and Suriname can be associated with category 1, whereas the expected problems in Ghana, Mali and Yemen can be associated with category 2.

The effects of the two risk categories are primarily:

- loss of production of agricultural land and more difficult housing conditions;
- water shortage for all uses, i.e. domestic, industrial and agricultural.

Table 3.5 provides an overview of the vulnerability profiles in the six countries investigated.

Higher river discharges

The studies on vulnerability of water resources to climate change indicate important variations in the runoff levels, depending on the climate scenarios considered and the basin studied and its location. This risk is typical for river basins with perennial rivers (rivers that flow year-round). Based on the studies described here, it can be concluded that under different scenarios for the same river basin, both increase and decrease in river runoff can occur. If this is the

---

[4] See Note 9, Chapter 1.
[5] DHV, PO Box 484, 3800 AL Amersfoort, The Netherlands. Tel.: +31 (0)33 468 22 10, Fax: +31 (0)33 4682301, Email: hans.leenen@dhv.nl
[6] IPCC (2001), Chapter 4 currently presents an excellent overview.

Table 3.5. Background on vulnerability profiles (Source: Dorland et al., 2001).

Country	Risk	Technical constraints	Legal/institutional constraints	Financial constraints
Bolivia	Higher incidence of floods and erosion particularly in the La Paz region	The technical levels of staff are reasonable to good. Further training is required to improve the capacity for complex assessments. Lack of reliable hydro/meteo and river flow data hamper the assessments of key pilot areas	The staff of the San Andres University were dedicated and able to learn fast in the field of water resources modelling. The capacity for multidisciplinary cooperation could be strengthened. The Secretariat of the PNCC functions well as a focal point for climate change in Bolivia. However, future funding is not assured	Physical adaptation measures can be implemented gradually, integrated in the design of new infrastructure, resulting in an increased cost level, probably requiring international funding
Ecuador	Risk of flooding due to increased rainfall and river discharge. The southern part of the Gulf is the most vulnerable area	Sufficiently skilled staff and a well-maintained hydrological database. A strong need for improving river basin management by improving transfer of relevant software and training of staff	Professional cooperation between the Ministry of Environment, the Hydrological Institute and local researchers	No information available
Ghana	Water scarcity is expected. The situation of hydropower generation is judged to be extremely vulnerable. Rainfed agriculture will require irrigation under conditions of climate change	Relevant experts, know-how and adequate data series are available for complex and multidisciplinary assessments of climate change impacts	The water resources team has succeeded well in involving a range of institutions and making experts in different disciplines contribute in the present study. Further development in management and coordination skills will be beneficial to future assessments. The recent establishment of the Water Resources Commission (WRC) can be instrumental in giving adequate weight to climate change concerns in future water resource planning	Physical adaptation measures can be implemented gradually, integrated in the design of new infrastructure, resulting in an increased cost level, probably requiring international funding

Continued

Table 3.5. Background on vulnerability profiles (Source: Dorland et al., 2001) (continued).

Country	Risk	Technical constraints	Legal/institutional constraints	Financial constraints
Mali	Risk of increase of water deficits in agricultural production, in particular for cotton, maize and millet (two river basins have been studied)	Basic technical skills are available but may need some polishing and further upgrading	No information	No information
Suriname	Suriname assessed the specific water resource risks in the coastal zone in terms of flooding and increased needs for drainage capacity	The availability of skilled and experienced personnel at all levels is a major problem	Staff are well motivated and knowledgeable, but total capacity is limited. Authorities in general are vertically structured. Cultural barriers hamper horizontal consultation and multidisciplinary cooperation	The replacement of the present gravity drainage by pumping facilities requires international funding, estimated at US$170 million
Yemen	Groundwater resources are insufficient under conditions of temperature increase and rainfall decrease. This will result in the long term in a limitation of further agricultural development due to lack of good quality water. Domestic and industrial water supplies are secured, although at higher cost due to a decline of the groundwater level	Limited capacity, university-based study team Technical skill at graduate level without relevant working experience	Focal point for climate change issues through Environmental Protection Council, including the ministries involved Project Management Unit for climate change management. The continuation of the coordination activities depends on international funding	No information available

case, the risks of increased flood levels will prevail in a river basin for which there is no water scarcity to begin with.

Erosion can be seen as a derived effect, although it is caused not only by flooding, but also by human interventions. Bolivia, Ecuador and Suriname are typical countries that run a higher flood risk. The country studies indicate that in Bolivia and Ecuador the risk is primarily related to higher river discharges, while in Suriname the flooding risk is related to sea level rise (SLR), causing a need for additional drainage capacity in the Nickery Delta area.

Increased risk of water scarcity

This risk is typical for the arid and semi-arid countries. The drier the climate, the more sensitive is the local hydrology. In dry climates, relatively small changes in temperature and precipitation may cause relatively large changes in runoff. Arid and semi-arid regions will therefore be particularly sensitive to reduced rainfall and to increased evaporation and plant transpiration. Many climate models project declining mean precipitation in the already dry regions of central Asia, the Mediterranean, southern Africa and Australia. Water scarcity is not only reflected in reduced surface water availability; it will also lead to reduced recharge of groundwater aquifers. Ghana, Yemen and Mali are typical countries that run a high risk of water scarcity.

3.4.2 Adaptation strategies

The specific country adaptation strategies in the six countries are presented in Table 3.6. In summary, adaptation strategies for the two listed risk categories comprise the following:

1. Adaptation to increased risk of flooding and erosion due to higher river discharges:

- increased drainage capacity;
- flood protection and other flood prevention works;
- erosion control through reforestation;
- changes in land use.

Combating and adapting to flood risk requires familiar technical interventions that are already applied elsewhere, or in the country itself. Adaptation would typically require the design and implementation of flood protection works, increased drainage capacity and changes in land use.

2. Adaptation to increased risk of water scarcity:

- more efficient water use through water and wastewater pricing policies;
- improve water availability throughout the year by reservoir management;
- introduction of water saving techniques;
- inter-basin water transfer;
- improved wastewater management;
- wastewater reuse in non-potable use, such as irrigation;

Table 3.6. Adaptation measures.

Country	Identified adaptation measures
Bolivia	Inter-basin water transfers, reduction of water losses from water conveyance and distribution systems, improved wastewater management, more efficient water use in all sectors, introduction of water saving techniques, water and wastewater pricing policies, erosion control (reforestation)
Ecuador	Adaptation measures were not elaborated
Ghana	Inter-basin water transfers, reduction of water losses from water conveyance and distribution systems, artificial recharge of groundwater to reduce evaporation, improved land use practices and irrigation control to reduce siltation of surface water reservoirs, more efficient water use in all sectors, recycling of water for non-potable uses, introduce cash crops with low water use, diversification of the sources for energy generation (move from hydropower to natural gas or other energy carriers), improved urban planning to avoid damage in areas that may become increasingly flood prone
Mali	Increase irrigation quantity through supply from river water or groundwater pumps, optimized regulation of river discharges through reservoir management, implement an early warning system to identify the risk of crop deficits, choose different crops in agreement with available water, improve the agrometeorological information available to the rural population, implement an agrometeorological database system for the farmers in rural areas, increase the mobility of populations in case of extreme droughts
Suriname	Pumping systems to guarantee adequate drainage of agricultural, urban and industrial land
Yemen	Water preservation, reduction of water losses, efficient water use, changes in crops

- adaptation of cropping patterns, switching to crops that consume less water;
- set up a detection and early warning system;
- increase the mobility of the population in case of extreme drought.

Water scarcity is not a new risk for many countries, but a phenomenon that is often already experienced in arid and semi-arid countries. For example depletion of groundwater resources is quite common. Climate change requires taking action in cases where water scarcity problems are already present. Yemen, where groundwater resources are rapidly depleting, is a typical example.

Strategies to combat water scarcity require not only technical interventions, but also institutional interventions related to water use management and water demand management.

Follow-up activities

The individual country's suggested follow-up activities are presented in Table 3.7. Ecuador has clearly indicated the necessity for evaluation of the economic values of water resources and ecosystems. Actually, this approach is extremely important to all countries, and should be elaborated further (see also Subsection 4.3.2).

3.5 Adaptation in Coastal Zones

Marcel Rozemeijer[7], Arjan van der Weck[8], Frank van der Meulen[9] and Rob Misdorp[10]

3.5.1 Introduction

This section presents the cross-country sector syntheses of the effects of SLR on the coastal

Table 3.7. Suggested follow-up activities by country.

Country	Follow-up activities
Bolivia	Incorporation of extreme climatic events (seasonal variation, extreme rainfall, peak floods and droughts) into future climate change projects
	Multidisciplinary evaluation of impacts and adaptation strategy
Ecuador	Expansion of current knowledge and experiences to carry out vulnerability studies in other river basins
	Make improvements to and optimize the hydrometeorological observation networks
	Assessment of the economic value of water resources and ecosystems
Ghana	Incorporation of seasonal variation and extreme events (rainfall, peak floods, droughts) into future climate change projects
	Adoption of international approach in Volta basin
	Assessment of the possibility of recharge to groundwater
Mali	Elaborating the study of anticipating optimal strategies in terms of mobilizing resources and possible assistance to affected populations
Suriname	Promotion of interdisciplinary cooperation
	Transfer of software models and training of staff
	Set up monitoring system
Yemen	Incorporation of vulnerability assessment results into Yemen's national development plan
	Extrapolation of pilot areas to national level
	Inventory of stakeholders and data for the formulation of an action plan for the water sector

---

[7] See Note 8, Chapter 1.
[8] See Note 9, Chapter 1.
[9] National Institute of Coastal and Marine Management, (Netherlands Ministry of Transport, Public Works and Water Management), PO Box 20907, 2500 EX Den Haag, The Netherlands. Tel.: +31 (0)70-3114311, Email: f.vdmeulen@rikz.rws.minvenw.nl
[10] For address and telephone number see Note 9. Email: r.misdorp@chello.nl

Table 3.8. Overview of the impacts of SLR using vulnerability classes as defined by the IPCC (1991, 1994).

Impact category	Vulnerability class			
	Low	Medium	High	Critical
Population at risk (% of total population)	<1	1–10	11–50	>50
Capital at risk (% of GNP)	<1	1–3	4–10	>10
Land at loss (% of total area)	<3	3–10	10–30	>30
Protection costs at risk (% of GNP)	<0.05	0.05–0.25	0.25–1	>1
Wetland at loss (% of total area)	<3	3–10	10–30	>30

zones of those countries financed by the NCCSAP. The present vulnerability of coastal areas, (owing to unsustainable use and coastal degradation) will be exacerbated by the expected SLR and other impacts of climate change. In 1990, the IPCC called for assessments of vulnerability to climate change. A common methodology for vulnerability assessment was developed (IPCC, 1991), to examine coastal vulnerabilities resulting from climate change impacts such as SLR (see also Subsection 1.5.2.).

A large number of vulnerability assessments for different coastal countries have been conducted. During the World Coast Conference (IPCC, 1994) – under the auspices of the IPCC – more than 40 coastal countries reported their progress on vulnerability assessments. Twelve countries reported their preliminary follow up on countrywide Integrated Coastal Zone Management (ICZM) efforts, while 11 countries reported on local ICZM area cases studies (IPCC, 1994).

The NCCSAP programme, encompassing 11 coastal countries, presented its results during the World Water Forum 2000 in The Hague (NCCSAP, 2000). The results of these 11 vulnerability assesments are reviewed here and the experiences with the application of the IPCC Common Methodology are examined in light of possible follow-up actions. Four vulnerability assessment studies in Bangladesh, Egypt, Nicaragua and Vietnam are included in the present review as they were also funded by the Netherlands Ministry of Foreign Affairs prior to the NCCSAP and for reasons of comparison and exemplary ICZM follow-up activities.

3.5.2 The impacts of SLR

This subsection shows that all the investigated countries will face severe problems as a consequence of SLR in the (near) future. Table 3.8 gives an overview of the quantification of effects according to the IPCC methodology (IPCC, 1991, 1994). Table 3.9 presents an overview of the countries, the areas studied, the climate change and socio-economic scenarios used in the assessments and the year in which most of the work in the studies was completed. The table illustrates the variety in geographic scope and the scenarios used. The country studies included qualitative and quantitative assessments where possible. In the quantitative assessments primary physical impacts of SLR on the population, the capital value and areas at risk were analysed. Estimates were given in absolute figures (i.e. number of people at risk, capital value at risk in dollars and areas at risk in km^2) or in relative figures (i.e. percentage of the country's/area's total population at risk). Capital value at risk is given as a percentage of the GDP or GNP or area at risk as percentage of the total area. In the IPCC terminology the following vulnerability classes are distinguished: low, medium, high, critical. See Table 3.8 for classes and impact categories.

Comparing Table 3.9 with Table 3.8 shows a complete range in effects. All NCCSAP countries fall into different impact categories and different vulnerability classes, ranging from 'low' to 'critical'. We will discuss some general aspects. We will not discuss these figures in further detail here, as comparison is difficult due to the different assessment levels and country situations.

The categories of people at risk and area at risk demonstrate a similar pattern of low and medium impact. Most countries are large, with a relatively small coastal zone that is poorly developed and sparsely populated. Only Bangladesh, Ecuador, Suriname and Vietnam show a large relative impact of SLR on the population, posing an impact on the whole society.

Table 3.9 also shows that most countries have a problem with the risk of losing large capital investments (capital value at risk: high to critical). Densely populated countries such as Bangladesh and Vietnam are at risk to lose from 10% to 80%, respectively. Even a country with a sparsely populated coast, such as Colombia, has a risk of losing 4.5% of the invested capital.

Another category with high to critical impact according to IPCC definitions is that of the wetlands at risk. All countries assessed will probably not be able to finance protection. One wonders to what extent the wetlands are really threatened. Probably the natural coastal dynamics could aid in preventing loss of wetland area. A prerequisite, however, is the availability of natural material such as sand and silt and a minimal flow of these materials along-shore and on-shore. In general, capital investments in the coastal zone (new or existing) tend to block or disturb these natural dynamics. Careful design of capital investments and/or occasional sand nourishment would be important.

Compared to other countries assessed with the vulnerability assessment methodology (IPCC, 1992a, 1994), the NCCSAP countries are different on one point. They show much less population at risk. They are comparable for the other aspects (capital value at risk, area at risk, wetlands at risk and protection costs (IPCC, 1992a, 1994)). Most NCCSAP countries have an underdeveloped coastal zone. This is contrary to the general pattern all over the world of highly occupied coastal zones. In this light, it is not advisable to treat the NCCSAP countries with the general strategies of the IPCC, but to carefully assess the actual situation, prognoses and forthcoming needs.

3.5.3 Response strategies in general

The current section discusses the potential strategies for the countries studied. It will become clear that, depending on the impact and country characteristics, different strategies are advisable.

There are three basic types of response strategies: protect, accommodate and retreat. It is not in the scope of the vulnerability assessment method to perform a full cost–benefit analysis of all kinds of response options. The vulnerability assessment approach is therefore based on a simplified procedure, only considering the most straightforward options. For example a full protection strategy is often not realistic for most locations where measures should be taken, given the economic, social, cultural, legal and institutional circumstances in the countries. Table 3.9 presents the estimated adaptation costs for combinations of strategies (see also Sections 2.3, 2.4, 2.5 and 2.11). An indicative quick overview of potential response strategies for NCCSAP countries in relation to impacts is given in the following paragraphs.

Low overall impact

Some countries have large unoccupied coastal zones. Although the results suggest that such countries, e.g. Colombia, Costa Rica, Nicaragua and Yemen, have a low overall vulnerability (considering the five aspects of Table 3.8), the vulnerability in specific locations (such as cities or other areas with high economic interests) can still be high to critical. It seems the best option is a combined approach, with: (i) protection and careful future development planning in high economic interest areas; and (ii) carefully planned accommodate and retreat measures in other high interest areas (such as nature reserves).

Critical relative impact, low absolute impact: retreat?

Some countries show high relative impact (expressed as a ratio of the entire country, Table 3.9 column 'Area at risk'). Expressed as part of the country's economic situation, the impact could be critical. Expressed in absolute figures the impact is less grave. Comparing absolute figures, SLR has low absolute impacts in terms of absolute numbers of people or money. Here a careful study

Table 3.9. An overview of areas studied, climate scenarios and economic scenarios used, identified risks of SLR without adaptation and the adaptation costs.

Country	Area studied	Climate change scenarios	Economic scenarios	Year[b]	People at risk	Capital at risk	Area at risk	Adaptation measures considered	Protection costs under scenarios studied[a]
Bangladesh	Total country (divided into nine vulnerability zones)	Null scenario, Moderate scenario climate changes (30 cm, 2°C by 2100) Severe climate changes (100 cm, 4°C by year 2100)	Business as usual 3.4% annual growth GDP up to 2010 High-development 5.7% annual growth GDP up to 2010	1994	About 40% of the population at risk of annual flooding in 2010 under severe climate change scenario	Up to 12% of GNP annually in 2010 with 100 cm SLR scenario	Up to 75% of land area is susceptible to annual flooding in 2010 with 100 cm SLR		No estimates
Colombia	Pacific coast Atlantic coast pilot areas	Null scenario 30 cm in 30 years 100 cm in 100 years	Optimistic: 4.5% annual growth GDP; 1.4% annual growth population Pessimistic: 3.5% annual growth GDP; 1.1% annual growth population	2003	±2% (1.7 million persons) 0.21% (2030) 0.51% (2100)	Up to 4.5% of GDP (US$137 million) 0.2% (2030) 0.7% (2100)	±7000 km² will be lost due to 1 m SLR	Protect Accommodate Retreat	From 3.4% of GDP in 2001 to 10% in 2100 (US$530 million in 2000 values)
Costa Rica	Pilot areas on the Pacific coast	Optimistic: 30 cm/100 years Pessimistic: 100 cm/100 years		1999	Over 30,000 people (2% population)	No estimates	?		No estimates
Ecuador	Country study pilot areas	Null scenario, 30 cm in 100 years, 100 cm in 100 years	3% annual growth GNP 2.2% annual growth population	1998	7–12% in 2010 at risk annually	13–23% at risk annually in 2010	32–41% for one in a 100-year flood	Total protection	US$2–4 billion (undiscounted)
Egypt	Mediterranean coast Red Sea coast Suez Canal area coast	Null scenario 30 cm in 30 years 100 cm in 100 years	Business as usual (4.6% annual growth GNP – 2.5% annual growth population)	1992	5.4 million people	Up to 14% of GNP	5,000 km² of land		5–10% of GNP

Cross-country Syntheses

Country	Study area	Scenarios	Socio-economic assumptions	Year	People affected	Economic impact	Land area impact	Response	Cost
Ghana	Country study pilot areas	Null scenario and 100 cm SLR in 2100	One socio-economic scenario, a defined response scenario and estimates of the associated cost	1998	±1% (132,000) people	No estimates	Two-thirds of the land area of the East Coast 0.5% of the country's land area		±US$590 million (1997 values)
Kazakhstan	Caspian Sea	Three sea levels of the Caspian Sea relative to MSL: −27 m −25 m −22 m	Not considered	1999	600,000 people (2% of the population of Kazakhstan)	No estimates	50 km inland on eastern coast of the Caspian Sea for −22 m scenario	Protection plan: Dyke construction Settlement relocation	±US$6.4 billion (undiscounted)
Nicaragua	Pacific coast Atlantic coast	Null scenario 100 cm in 100 years	?	1995	?	?	±6,000 km²		No estimates
Senegal	Two case study areas	Null scenario Low: 7 cm to 2050 and up to 20 cm in 2100 Medium: 20 cm in 2050 and 50 cm in 2100 High: 40 cm in 2050 and 86 cm in 2100	2–3% population growth rate 0.4% agricultural production growth rate 3% and 6% discount rates	1999	±180,000 people (2% of the population) for 40 cm SLR in 2050 ±1.6 million people (14% of the population) for 86 cm SLR in 2100	US$490 million in 2050 with a 3% discount rate	948 km² (0.5% of the total area of the country)	Protection	±US$250 million (protection and replanting of dune areas) for 40 cm SLR in 2050 with a 3% discount rate
Suriname	Country study Atlantic coast All river basins	Null scenario 30 cm in 30 years including 10% precipitation increase	5% annual growth of GNP up to 2025 1.2% annual increase in population	1999	±70,000 people (representing 11% of the national population) will be at risk in the 30 cm scenario	±US$240 million 11% of the annual GDP will be at risk in the 30 cm scenario	±4,000 km², especially the capital is threatened (95% of the total area)		±US$470 million (1999 values)
Vietnam	Country study pilot areas	Null scenario 30 cm in 30 years 100 cm in 100 years	5–10% annual growth GDP up to 2025; difference between mountain and coastal provinces	1996	±17 million people due to annual flooding	±US$17 billion (80% of GNP) due to annual flooding In 30 years US$270 billion (125% of GNP)	±1750 km² (60% of Vietnam's coastal wetlands)		±US$9 billion (40% of GNP) (undiscounted)
Yemen	One pilot area including Hodeidah city	Null scenario 100 cm in 100 years		2000	±91,000 people	±US$1.3 billion	±50 km²		±US$20 million (1999 values)

[a] Unless stated otherwise.
[b] The year in which most of the work on the study was completed.

of the response strategy should be made before acting. Ghana and Suriname represent countries where protection measures are not the first issues to think about. In most cases, small numbers of people are at risk. Retreating the major investments and big cities might be a good option given the small absolute number of people and the small absolute contribution to GNP. It would be interesting to study how natural dynamics of the coast could contribute to protection and make the coast more resilient. A combination of importing material (e.g. sand nourishment) and giving space to the natural processes of erosion and accumulation (no hard construction in the coastal zone) might give protection in the long run without requiring major investments.

Critical impact but protection possible

For some countries, protection measures seem affordable. The cost–benefit ratio turns out to be advantageous for protection measures. Senegal is such an example. A conservative estimate of the potential effect of SLR on the GNP shows a critical impact. A protection strategy would consist of sand nourishment (using natural processes and materials). Taking the relatively low costs of such protection into account in relation to the high benefits, a strategy of protection seems a realistic and likely option.

Critical impact and high protection costs

Some countries are in a very critical position: the coastal zone offers a major contribution to the GNP and will be seriously affected by SLR. On the other hand, full protection measures would lead to a pay back time of more than 1000 years. The large investment costs for protection are not realistic in relation to the potential benefits. A strategy of accommodation and retreat seems the best option even though the societal impact is high. This seems preferable over the high potential costs. The use of the natural dynamics of the coast in developing response strategies is recommended, also because this will be cheaper. We encounter a potential situation like this in Ecuador.

Critical impact and combined strategies

The most hazardous situation is in countries where the potential impact of SLR is critical and where large areas, large parts of the population (millions of people) and substantial capital values are at risk. Considering the large societal impact, retreat or accommodation is hardly acceptable. Often it is impossible to reallocate people and capital investments on such a scale. On the other hand, the costs of full protection also exceed any potential profit or yield of investments on such a large scale. Bangladesh, Egypt and Vietnam represent this situation (Table 3.9). Large-scale protection combined with accommodation and retreat seems to be both required and the only option, but this would require huge financial resources.

3.5.4 Action plans defined in the vulnerability assessment

This section gives an overview of the action plans defined in the vulnerability assessment. As part of the country vulnerability assessment recommendations for the future, action plans were formulated in most of the vulnerability assessment studies. Table 3.10 presents an overview of actions mentioned in the action plans, as seen during the years of study. The countries have not mentioned economic and financial aspects. It is clear that for all of the countries considered, extensive follow-up activities are necessary in order to cope with the consequences of SLR.

In some countries, physical (engineering) solutions (both hard measures such as groins and soft measures such as sand nourishment) are proposed, for instance in the cases of Senegal, Vietnam, Egypt and Yemen. In almost all cases, additional research is recommended to gather more data and information before the implementation of specific solutions. In view of the character and scale of the vulnerability assessment studies (a first assessment, often on a nationwide scale) this seems to be a sensible way of advancing towards sustainable and sound solutions to the problems most countries will face in the (near) future.

In almost all countries, the introduction of ICZM is taken as the most adequate way of

Table 3.10. Overview of actions mentioned in the action plans for each country.

Actions mentioned in action plans	Ba	Co	CR	Ec	Eg	Gh	Ka	Ni	Se	Su	Vi	Ye
Legislative, institutional and organizational aspects												
Existence of an organizational structure												
Awareness raising at decision-making level		x					x	x		x	x	x
Horizontal integration	x	x										
High level ICZM coordination	x				x				x	x	x	
Integrating ICZM coordination with regular departments	x	x									x	
Coordination between departments		x			x				x			
(International) river basin management	x											
Operational level of structures												
Vertical integration		x						x			x	
Public participation	x							x		x		
Legal instruments/enforcement				x				x		x		
ICZM plan/centre									x	x	x	x
Integrating spatial planning with coastal planning				x								
Platforms for negotiations of rights and needs	x											
Nature conservation regulations			x									
Pilot projects										x		
Capabilities of personnel												
Capacity building	x	x		x	x	x		x		x	x	x
Technical aspects												
Existence of an organizational structure												
(International) knowledge transfer	x											
Research on coastal processes		x		x	x	x	x	x	x		x	x
Operational level of structures												
Design criteria for coastal structures		x			x							
Capabilities of personnel												
Capacity building	x	x		x				x		x	x	
Organizing data collection		x		x		x	x	x		x	x	x
Infrastructures for:												
Coastal protection							x	x		x		
Irrigation					x							
Drainage					x				x			
Prevention of saline intrusion					x				x	x		
Protection of investments								x				
Reallocation of infrastructures								x				
Cultural and social aspects												
Reallocation of population								x				
Operational level of structures												
Public participation		x							x	x		
Awareness raising			x						x	x	x	

[a]Ba, Bangladesh; Co, Colombia; CR, Costa Rica; Ec, Ecuador; Eg, Egypt; Gh, Ghana; Ka, Kazakhstan; Ni, Nicaragua; Se, Senegal; Su, Suriname; Vi, Vietnam; Ye, Yemen.

dealing with future consequences of SLR. Adopting ICZM involves aspects in the institutional, organizational and legislative public administration and governance, such as: (i) stakeholder analysis and participation; (ii) horizontal integration of disciplines (involving various disciplines and governance authorities); (iii) vertical integration (involving the various levels of government administration from national to municipal); and (iv) raising awareness in the civilian population at risk (e.g. through local pilot projects). In many cases, the present knowledge of the natural and the socio-economic coastal system is seen as being too limited. Data collection and management, education, training and transfer of tools are therefore often mentioned as important conditions for the successful introduction of ICZM.

3.5.5 Feasibility of implementation of action plans

The current section discusses the feasibility of implementation of the action plans. Most countries will need substantial external funding for the implementation of response strategies.

Table 3.11 presents the feasibility of the defined action plans. For all four aspects of society (Table 3.11) a qualitative estimate was made using a scale of 'no problem', 'partial problem' to 'problem'. Three levels of organization were used: (A) Existence of basic structures; (B) Performance/operational level; (C) Capability of performers. The more problems encountered at level A, the less feasible the action is. The evaluation classifications are:

- *Problem*: this means that actions are not readily feasible, or not feasible without extensive (international) assistance. The existence of the basic organizational structures is not guaranteed. For the legislative structures this could mean, e.g., extensive reorganization. For economic aspects, *problem* could mean a requirement for extensive investment that is not readily available;
- *Partial problem*: the actions can be implemented. Problems occur more on the operational level. Initiatives can be implemented on a national scale, though at the local level problems will be encountered;
- *No problem*: problems are mainly at the capability level of the performers. Actions can be relatively easily implemented at the local level with little or no (international) assistance.

Table 3.11 Feasibility of implementation of the action plans.

Country	Vulnerability assessment			
	LOI[a]	ECF[b]	TEC[c]	CSO[d]
Bangladesh	problem	problem	partial problem	problem
Colombia	partial problem	problem	partial problem	problem
Costa Rica	no problem	na[e]	no problem	na
Ecuador	partial problem	problem	no problem	na
Egypt	partial problem	problem	problem	no problem
Ghana	partial problem	problem	partial problem	na
Kazakhstan	no problem	problem	no problem	na
Nicaragua	problem	problem	partial problem	partial problem
Senegal	partial problem	problem	partial problem	partial problem
Suriname	partial problem	problem	no problem	partial problem
Vietnam	problem	problem	partial problem	partial problem
Yemen	partial problem	partial problem	problem	na

[a]LOI, Legislative, institutional and organizational aspects.
[b]ECF, Economic and financial aspects.
[c]TEC, Technical aspects.
[d]CSO, Cultural and social aspects.
[e]na, Not analysed.

Note that some studies were carried out earlier than others (Table 3.9). Table 3.11 therefore does not represent the present situation for all cases. It can be concluded from Table 3.11 that economic and financial circumstances are a serious barrier to implementing action plans and response strategies in all countries. This is due to the large negative impacts of SLR. High costs are associated with adaptation and huge financial resources are needed.

Institutional, organizational or legislative circumstances are also often seen as an important barrier to implementation of response strategies (see also NCCSAP, 2000; Dorland et al., 2001). It can be concluded that only a few countries, i.e. Costa Rica, Kazakhstan, Ecuador, Egypt and to a lesser extent Suriname, have an ICZM system that enables participation of all stakeholders (including the general public) (Table 3.11; NCCSAP, 2000; Dorland et al., 2001). The other countries seem to lack a coordinated ICZM. In several countries, the centralized government and central planning systems were blocking the appropriate nesting of ICZM in all layers of public administration.

Technical aspects are also mentioned as (partially) problematic in several countries. These problems are mostly related to knowledge. Almost all countries need to organize data collection on the coastal system, hand in hand with capacity building of local staff and research on coastal processes (Table 3.11).

Cultural and social aspects are seen as the least problematic, although it should be noted that these aspects received less attention than the other aspects in most studies and hence need more detailed assessment. Cultural and social aspects include three levels. The first level deals with cultural and political stability in the country. Are there any ethnic or behavioural differences in the society that lead to unstable situations? The other aspects measured deal more with institutional aspects of being able to organize stakeholders. In this sense, the first aspect deals more with the (cultural) interaction between people and the other two deal with the organization of the interaction. A lack of sufficient stakeholder involvement was mentioned as one of the main problems in many countries. It is a serious barrier to the implementation of ICZM. Colombia and Yemen also experience internal social conflicts that potentially block ICZM.

3.6 Adaptation and Land Use

Jan Verhagen[11] and Frits Mohren[12]

3.6.1 Agriculture

The methodology

Climate change and climate variability can substantially affect agricultural production and have a dramatic impact on local and national economies. Vulnerability of the agricultural sector to environmental change is, however, a complex issue. Most studies focused on the direct impact of climate change on crop production levels. Possible effects on farm management and socio-economic and socio-cultural structures were hardly considered. Isolating the effect of climate change on a complex system is not a trivial matter. The IVM/UNEP *Handbook on Methods for Climate Change Impact Assessment and Adaptation Strategies* (Feenstra et al., 1998) provided to the teams emphasizes the importance of an integrated assessment. However, it does not provide the tools to perform this complex task.

Simulation models are not tools that can be used off the shelf. Understanding and working with complex dynamic simulation models and understanding and interpreting the results requires not only understanding of the system that is modelled but also skills to evaluate whether the translation of the system behaviour to the model was done correctly. It is obvious that these requirements cannot be met in a short period. It does not matter whether one wants to use models for scientific purposes (models to understand processes) or for engineering (models to assess system behaviour). A basic understanding is required.

---

[11] See Note 5, Chapter 1.
[12] See Note 7, Chapter 1.

The NCCSAP studies' models were used in an exploratory manner, providing information on the direction of change and not so much on the exact magnitude of this change. The full range of options, from detailed crop and livestock models to rating systems based on expert judgement, was used.

For most countries crop yields are expected to decrease mainly as a result of decreasing precipitation levels. Irrigation is an option in some cases but because the agricultural study was not linked to the water sector study, the implications for the landscape and watershed were not assessed. Selection of new crops and adjustments in soil management were also identified as possible adaptation options.

In cases where lack of data and experience with dynamic simulation models were limiting, an intermediate quantitative model/framework was proposed. This semi-quantitative approach makes use of both knowledge of basic phenological, physiological, morphological and physical processes, and expert judgement. It follows the logic of most dynamic crop simulation models but is based on a different level of detail. In this approach crop production is quantified using basic rules. Several production levels are defined. The analysis starts from a non-limiting production level (attainable production, determined only by crop characteristics, radiation and temperature), followed by the introduction of various yield-limiting and -reducing factors. The production levels are:

1. Potential production.
2. Yield reduction as a result of water stress.
3. Yield reduction because of lack of nutrients.
4. Yield reduction due to weed competition.
5. Yield reduction due to diseases.
6. Yield reduction as a result of pest incidence.

The first level is calculated using knowledge on crop phenology and growth rates under the prevailing environmental conditions, and requires information on incoming solar radiation and temperature. The subsequent levels of reduction are quantified based on literature and expert knowledge. The first reduction level (and to a lesser extent the second level) can be made spatially explicit using soil information. Other levels can be related to management, crop variety and climate.

This approach makes optimal use of both quantitative insights in the various processes and expert knowledge, and moreover it is in balance with the available data.

The effects of climate change are related to three components:

- change in precipitation – quantity as well as spatial and temporal distribution;
- changes in CO_2 level;
- changes in temperature.

Production levels

The effects of changes in climate on production level can be estimated, using the same approach. Changes in precipitation will modify the magnitude of the yield reduction due to water stress, changes in CO_2 level will affect the growth rate, changes in temperature will influence the length of the phenological stages. When evaluating the impact, the effects of each individual factor and the possible interactions and feedback mechanisms should be considered.

Also adaptations to mitigate possible negative effects can be defined in terms relevant to the semi-quantitative approach: introduction of varieties with a different phenological pattern, use of fertilizer to optimize the efficiency of use of (scarce) water resources, etc.

Results

Some results of the individual country studies can be found in Chapter 2. In this section an overview of tools and methodologies is given.

Most studies aimed at linking food production to food security issues, combining climate change with demographic development, technological options and socio-economic constraints.

Arable farming, for both commercial production and local consumption, was studied; only the Mongolian and Bolivian studies included livestock. When moving up from field to the farm and regional scales, it is in most countries imperative to include livestock. Table 3.12 shows the sectors included in the study and in Table 3.13 the range of crops is presented.

Table 3.12. Sectors included in the study.

Country	Arable crops	Grassland	Livestock
Costa Rica	+		
Bhutan	+		
Bolivia	+	+	+
Mali	+		
Senegal	+		
Yemen	+		

Table 3.13. Crops studied by the agricultural teams.

Crop	Costa Rica	Bhutan	Bolivia	Mali	Senegal	Yemen
Coffee	+					
Cotton				+		
Potato			+			+
Wheat	+					+
Rice		+			+	
Groundnut					+	
Millet					+	
Maize					+	
Sorghum					+	
Dry bean			+			
Oats						+

cases the documentation and underlying principles of these models were examined by the consultant. The duration and nature of the programme did not allow for experimentation, model development or thorough testing of the models. The models used by the agricultural teams are shown in Table 3.14.

The scaling up to the regional and national level was done in a qualitative manner. The link with other sectors such as water and forestry was done after completion of the individual studies by the national coordinator in cooperation with the Institute for Environmental Studies (IVM). Most countries participating in the NCCSAP already participated in the US CSP or other international programmes and wanted to build on this earlier study. However, this was not always possible, as the expertise acquired during these programmes was not directly available.

All studies started with the intention to use crop growth simulation models to assess the impact of climate change on crop production. DSSAT was clearly the most favoured model; this model was also used in the US Country Studies Program (US CSP). Experience with the use of DSSAT was not always readily available and alternative tools needed to be identified. For most situations locally or regionally developed models were available, in these

Table 3.14. Models used by the agricultural teams.

Model name	Bolivia	Bhutan	Costa Rica	Senegal	Mali	Yemen
DSSAT	+	+	+			
SPUR	+					
COFEA			+			
ARABHY				+		
SARRA				+		
Cotton					+	
ALES						+
Semi-quantitative						+
Qualitative						+

The process

Technical assistance works in well-defined studies when procedures and methods are clear and a basic knowledge of methods and tools is available in the study team. In most of the studies, it was clear that experience with modelling was rudimentary or lacking. Nevertheless most teams were very keen on using models. The 'learning by doing' process is rooted in the eagerness of the teams to work with models and to keep up with the latest methods and technologies. Obviously, there is a tension between the thirst for new tools and knowledge and the programme that was designed to feed the National Communications.

Balancing the ambition levels of the teams and keeping the output objective of the studies in focus turned out to be a major task of the technical assistance. Including the technical expertise in the first phases of project definition and including training programmes in the inception of the country studies overcame part of this problem.

Both the receiving and the delivering end of the technical assistance had to find a *modus operandi* to work together. The on-demand technical assistance was a new concept for most participants, roles and expectation patterns were not always clear. A complicating factor was that the relatively complex organizational structure of the programme sometimes led to miscommunications. The study teams did not work on stand-alone projects, but were part of multi-sectoral analysis, which, in most countries, was coordinated by the ministry of environment. The study teams for agriculture and forestry were typically part of the ministry of agriculture or agricultural universities.

The sector approach did not result in an integration of the various sectors, although clear links exist. Water and agriculture were discussed in isolation and adaptation options were sector based and limited to technological choices. Synergies and options at higher levels were not assessed. The current socio-economic context is essential in understanding food security and although the effects of climate change on socio-economic aspects of the agricultural system are difficult to assess, this context is crucial in defining adaptation options that aim for food security.

Ambition levels of the teams were high and in most cases this had to be moderated to safeguard the progress of the country studies. Training in the use of tools, such as crop simulation models, helped the teams to achieve some of their ambitions. It took time to adjust to the on-demand assistance, as it was new to most groups. Basically, all problems could be solved via intensive communication and regular visits.

Most studies spent a large amount of time determining the impacts of climate change on food and fibre production. After selecting an important food or cash crop, the studies focused on a simulation study of the effect of climate change on potential production levels including a scenario when water was limited. Although the studies yielded important and usable results, the development and political context was lacking. Integral sector based vulnerability studies were not conducted due to inadequate time and capacity. When the studies were conducted, assessing the possible impacts of climate change was a very important signal to society and policymakers. The work done in the NCCSAP provided a sound basis for future work on climate change impacts and vulnerabilities.

3.6.2 Forestry

The first point that arises when comparing the two projects in Bolivia and Costa Rica is the difference in size of the countries. Since Bolivia is more than 20 times larger than Costa Rica, the resources and efforts needed to reach an equivalent level of reliable information are substantially different. This is reflected in the results, i.e. in general, a more detailed analysis was made in Costa Rica. Nevertheless, both countries have mobilized several researchers from different disciplines as required for climate change studies. Both countries applied methodologies approved by the IPCC. The content of both studies is aimed at similar goals, despite the fact that the work was developed and integrated in different ways. Our contribution is focused on the forestry sector, but we also include a broader view of other areas of climate change impact analysis.

In the case of Bolivia, the forestry sector study contributed to three wider areas of analysis:

1. Emission inventories of GHGs in Bolivia, 1994.
2. Evaluation of mitigation options of GHG effects.
3. Ecosystem vulnerability to climate change.

In the case of Costa Rica, the main contribution of the forestry sector study was focused on the vulnerability of forests in Costa Rica to climate change. Nevertheless, this core study included several issues that were analysed and presented with more detail in other additional studies, presented as annexes. The number of different technical reports reflects the level of detail developed in the framework of the project.

A prime component of any climate change study is the description of the present climate and how this may change under certain future scenarios. A number of problems with this analysis arose in Bolivia due to the size and heterogeneity of the country. Global Circulation Model (GCM) scenarios were used for this analysis for both Bolivia and Costa Rica. These models were tested against actual data from meteorological stations located in each country (six in Costa Rica and 32 in Bolivia). Different models were analysed and checked. Those provided by the Hadley Centre appeared to be the most suitable for both countries but others were tested as well. Once the GCM scenarios were calibrated and tested, their results were entered into the MAGICC model. This software allows the user to determine changes in GHG concentrations, global mean surface air temperature and sea level resulting from anthropogenic emissions of GHGs (Wigley, 2004).

The emission scenario results generated by MAGICC were entered into the SCENGEN model. SCENGEN constructs a range of geographically explicit climate change scenarios for the world by exploiting the results from MAGICC and a set of GCM experiments, and combining these with observed global and regional climate data sets (Wigley, 2004). Three scenarios (optimistic, pessimistic and moderate) were developed for each country. For the purposes of the analysis, each country was divided into a number of geographical areas. Four areas were defined in Costa Rica, but only three were analysed. In Bolivia five areas were considered, but only four were analysed. Given the differences in the size of the two countries, it is evident that less detailed information is represented in the Bolivian analysis relative to that of Costa Rica.

Impact of climate change on forests

In order to relate climate change to changes of vegetation distribution, both the Bolivian and the Costa Rican studies made use of Holdridge's system of Life Zones classification (Holdridge et al., 1971). This classification relates the potential presence of a given forest type to the level of rainfall, bio-temperature and of potential evapotranspiration. In other words it relates equilibrium vegetation type to climatic conditions. The idea behind its application is to give evidence of how new scenarios may indicate future environmental restrictions to the actual vegetation distribution. Additionally, other important variables can be added to improve the analysis. For example, Costa Rica added information on dry months and soils to get a more detailed description and thus produced more detailed maps of potential vegetation cover-types.

The next step was to relate those maps to actual forest cover and to the presence of specific forest types, i.e. actual location, vegetation type and spread. In the case of Costa Rica, field measurements were carried out at 68 sites to quantify carbon stocks in a given specific forest type, as well as to verify the forest type in each location. Although only primary forest was considered, this effort is very valuable because it introduces new data, obtained with a reliable methodology. The maps developed within the present study were based on a scale of resolution of 1:200,000 and were also compared with the maps produced by Bolaños and Watson (1993) with the same spatial resolution.

The spatial resolution of the vegetation maps generated in Bolivia is 1:10,000,000. This low resolution was imposed by the limited spatial resolution of the meteorological data.

As data from meteorological stations were too scarce to cover the country, this information was not used to produce Holdridge's Life Zones distribution maps. Meteorological data, with a regular grid of 0.5°, provided by the International Institute for Applied Systems Analysis (IIASA) in Laxenburg, Austria, were used instead. The outputs of this model were compared with more detailed maps (1:1,000,000) elaborated between 1971 and 1975. These maps were somewhat supported with field verifications. The differences between maps probably result from the change of resolution. It is necessary, however, to obtain a better resolution and field verification in order to ensure that the actual representation of forests is feasible.

The Costa Rican study evaluated the vulnerability of flora and fauna based on the existence of their habitat. In Bolivia a gap model named GAP was applied to determine the effects of climate change on the production of tree species, which is interesting from an economic point of view. The limitation of this model application, as well as of other potentially applicable process-based models, is the availability of measured parameters needed to run such a model. Nevertheless, it is worthwhile to perform some specific analysis in areas where the information may be available. Complementary studies may also be helpful.

The contribution of the forestry sector to GHG emissions, i.e. changes in land and forestry use, was also analysed in both countries using land use models and future land use scenarios. The model used in Bolivia was COPATH3. This model is based on spreadsheet calculations of changing areas of land cover types. In the case of Costa Rica, several models like COPATH3 were checked. Finally, a model of deforestation rates was used. This model considers different variables related to deforestation, including institutional, physical and socio-economical aspects.

3.7 National Communications

Michiel van Drunen[13]

3.7.1 Introduction

National Communications to the United Nations Framework Convention on Climate Change (UNFCCC) have become not only one of the central features of the Convention process and of the active involvement of the Parties therein, but they are also one of the most important tools for bringing climate change concerns to the attention of policy-makers at the national level (UNFCCC, 2004a).

Originally, the guidelines for preparing the National Communications were not very detailed. The National Communications Support Programme (NCSP) provided a four-page 'Tentative list of technical issues for consideration' for non-Annex I Parties, which was based on the results of the second Conference of Parties (COP) in 1996.[14] The scope of the National Communications according to these guidelines is summarized in Box 3.1. Most of the guidelines provided by the NCSP referred to the emission inventory (one page) and the general description of steps (two pages).

National Communications and GHG inventories from Annex I Parties are subject to in-depth review by teams of independent experts. The aim is to provide a thorough technical assessment of each Party's commitments and steps taken towards their implementation. Periodic in-depth reviews of National Communications started in 1995. They typically draw on findings from

---

[13] See Note 1, Chapter 1.
[14] Point 9 of the Ministerial Declaration states that the Ministers and other heads of delegations present at the second session of the Conference of the Parties to the United Nations Framework Convention on Climate Change 'Welcome the efforts of developing country Parties to implement the Convention and thus to address climate change and its adverse impacts and, to this end, to make their initial national communications in accordance with guidelines adopted by the Conference of the Parties at its second session; and call on the GEF [Global Environment Facility] to provide expeditious and timely support to these Parties and initiate work towards a full replenishment in 1997.'

> **Box 3.1.** Scope of National Communications according to the 'Tentative list of technical issues for consideration' of the National Communications Support Programme (NCSP).
>
> The non-Annex I initial communications 'should include' three main components:
>
> - Greenhouse gas (GHG) inventory
> - General description of steps
> - Other information.
>
> Of these, the inventory is the key component and the only one for which non-Annex I Parties are required to follow explicit reporting guidelines. Technical assessment could therefore focus on the inventory.
>
> Under the general description of steps, various subcomponents can be included because the COP guidelines state that non-Annex I Parties 'should seek to include' these subcomponents 'as appropriate'. The key subcomponents are likely to be abatement, vulnerability, adaptation and systematic observation. Because of the flexibility in the COP guidelines, assessment of these subcomponents could be optional. However, some non-Annex I Parties may choose to have these subcomponents assessed if they are already included in the National Communications. An exception may be abatement, as countries are often sensitive about this issue.
>
> Similarly, 'other information' may include a list of projects for financing. Again, the COP guidelines state that this information can be provided on a 'voluntary basis', but many non-Annex I Parties may also wish to include this component in the assessment for the reasons stated above. The same would apply to 'Financial and technological needs and constraints', which are also part of the COP guidelines.

visits to the country concerned as well as desk-based studies. The guidelines of Box 3.1 were used by the reviewers hired by the NCSP to review the draft National Communications.

Non-Annex I Parties have gained much experience in their first round of National Communications. Building on this, the COP, at its eighth session, adopted a set of new and improved guidelines to help them prepare their second and, where appropriate, first and third National Communications. In 2003 the UNFCCC issued its user manual for the guidelines on National Communications from non-Annex I Parties (UNFCCC, 2004a), which was obviously not applicable to the studies described here. Its main features are summarized in Section 4.4.

National Communications were prepared as an integral part of, or in parallel to, the NCCSAP activities in Bolivia, Costa Rica, Ghana, Mongolia, Senegal and Yemen. In most cases, the results of the emission inventory and the vulnerability and adaptation studies were summarized in the National Communications.

3.7.2 Bolivia

Bolivia published its *First National Communication* to the UNFCCC in 2000 (Ministry of Sustainable Development and Planning, 2000). The English version is 138 pages long. Besides the introduction and the executive summary, it has five chapters: National Circumstances (20 pp.); National Greenhouse Gas Inventory (17 pp.); Vulnerability and Adaptation (34 pp.); Projections, Plans and Measures (30 pp., including 14 pages of mitigation plans); and Systematic Observation, Education and Public Awareness (17 pp.).

Bolivia's National Communication is well illustrated and provides a good overview of the GHG emissions, the vulnerability towards climate change and Bolivia's climate change policies. The NCCSAP supported the studies for the GHG inventory and the vulnerability and adaptation chapter.

The most important conclusions are:

- Activities related to land use change and forestry are the most important sources of

GHG emissions with 38.6 million t of CO_2, followed by the energy sector with 7.6 million t of CO_2. The total GHG emissions amount to 61.1 million t of CO_2 equivalents.
- Studies on vulnerability to climate change indicate that a probable 2°C temperature increase would not seriously damage cultivated areas if this increase goes together with precipitation increases. In the high plains, these conditions would be favourable for growing crops if provided with adaptation measures such as irrigation systems and improved cultural practices.
- Climate change in Bolivia affects human health directly and indirectly. The most frequent direct effects are: floods (Santa Cruz), landslides (La Paz), forest fires (Guarayos – Santa Cruz) and storms (Cochabamba), all of them increasing population mortality. Indirect effects include an increased occurrence of malaria and leishmaniasis.
- Cost-effective mitigation measures mainly include energy efficiency measures in the energy and industry sectors and reforestation.

3.7.3 Costa Rica

Costa Rica's Initial National Communication (República de Costa Rica, 2000) is written in Spanish, but it includes an executive summary in English. The other chapters are: National Circumstances (10 pp.); Legislation (9 pp.); Greenhouse Gas Emission Inventory (17 pp.); Climate Change Vulnerability (28 pp.); Mitigation Options (34 pp.); Related Programmes Concerning Sustainable Development, Systematic Research, Education, Awareness Raising and Capacity Building (6 pp.); and Financial and Technical Assistance (2 pp.).

The last chapter indicates that Costa Rica explicitly uses its National Communication to find financial aid for new climate related projects. The NCCSAP contributed to the vulnerability and adaptation sections.

The main conclusions include:

- The main contributions to Costa Rica's GHG emissions come from the energy sector (4.3 million t CO_2), industrial processes (0.4 million t) and agriculture (0.15 million t). Land use change has a net fixation of –0.9 million t CO_2.
- Studies show that there are yield reductions in dry-land rice as precipitation decreases. High temperatures also provoke considerable yield drops. Both potatoes and beans show an important decrease in yields with increasing temperature and precipitation variations. The effect of meteorological parameters on coffee yields varies and depends on water availability during the crop cycle. Increased CO_2 concentrations will have a positive effect on all crops investigated.
- The three climate scenarios show that the tropical and montane life zones seem to diminish while the premontane tends to increase. Rainforest life zones diminish in all levels. Dry, moist and wet tropical forests diminish significantly. But premontane moist and wet forests as well as lower montane moist forests expand.
- Mitigation options were identified in the transport sector (railroads, integrated traffic systems), the energy sector (renewables, energy conservation), the industrial sector (cement) and by reforestation and improved waste management.

3.7.4 Ghana

The *Initial National Communication of Ghana* (Environmental Protection Agency, 2001) includes the following chapters: Executive Summary (17 pp.); National Circumstances (13 pp.); Inventory of Greenhouse Gases (12 pp.); Impacts, Vulnerability and Adaptation (36 pp.); Measures Contributing to Addressing Climate Change (14 pp.); Sustainable Development and Planning (12 pp.); Research and Systematic Observation (9 pp.); Education, Training and Public Awareness (4 pp.); International Cooperation (2 pp.); and Proposed Climate Change Projects (42 pp.).

Ghana, like Costa Rica, uses its National Communication to find financial aid for new climate related projects. The NCCSAP contributed to the vulnerability and adaptation sections. In addition, the NCCSAP management reviewed the draft version of the National Communication.

Its main conclusions are:

- Carbon sinks in forested and afforested lands offset the total CO_2 emissions, thus making the country a net CO_2 remover (–0.4 million t). Although the total CH_4 emissions are lower than CO_2 emissions, the CO_2 equivalent emissions of CH_4 are about two to three times higher than the CO_2 emissions.
- The climate projections indicate that the average maximum temperature will increase by 2.5–3°C by 2100. The mean annual rainfall would decrease by 170 mm in the Sudan Savannah Zone, 74 mm in the Guinea Savannah Zone and 99 mm in the semi-deciduous Rainforest Zone, respectively by the year 2100. In the High Rainforest Zone, however, the mean annual rainfall was projected to increase by 1105 mm by the year 2100.
- Using the climate scenarios, it was projected that the maize yield would decrease in the Transition Zone from 0.5% in the year 2000 to 6.9% in the year 2020. The yield of millet, however, will not be affected by the projected climate change because millet is more drought tolerant.
- Simulations using projected climate change scenarios indicate reduction in flows between 30 and 40% for the year 2050 in all Ghana's river basins. Hydropower generation and irrigation are projected to be seriously affected.
- A preliminary assessment of the impacts of SLR shows that about two-thirds of the total land area potentially at risk of flooding and shoreline recession lies within the East Coast. A total of 1110 km^2 of land area may be lost as a consequence of a 1 m rise of sea level. The population at risk is estimated at 132,200.
- The loss of land by erosion and inundation will translate into loss of coastal habitats including important wetlands mostly in the Volta Delta. Increasing water depths and salinization of lagoons as a result of SLR will adversely impact the feeding success of migratory and resident birds.
- Estimated cost of constructing sea walls to protect all at-risk shorelines with populations greater than 10%/km^2 is US$1144 million. The protection of only important areas reduces that cost to US$590 million.
- Cost-effective mitigation measures include forest protection, replacement of biomass with LPG and gradual penetration of PV electricity and biogas.

3.7.5 Mongolia

The *Initial National Communication of Mongolia* (Institute of Meteorology and Hydrology, 2001) includes seven chapters: National Circumstances (11 pp.); National Greenhouse Gas Emissions (12 pp.); Greenhouse Gas Mitigation Issues (19 pp.); Climate Change, its Impacts and Adaptation Measures (29 pp.); Climate Change Response Policy (10 pp.); Research and Systematic Observation (3 pp.); and Education, Public Awareness and International Activity (4 pp.). The NCCSAP contributed to the mitigation chapter and the impacts and adaptation sections on agriculture.

The main conclusions of Mongolia's National Communication are:

- Fossil fuel combustion is the largest source of CO_2 emissions in Mongolia, accounting for about 60% of all emissions. The second largest source is from the conversion of grasslands for cultivation (20–27%). The total emissions of CO_2 in Mongolia reached 9.1 million t of CO_2 in 1994.
- Cost-effective mitigation options include Clean Coal Technology (effective dewatering systems, coal washing plants, selective mining and rock separation, briquetting technology); good housekeeping[15] of electricity and heat production, energy management, steam-saving technology and dry processing of cement in the industrial sector; and installation of thermostat radiator valves and balancing valves, improvement of building insulation, installation of

---

[15] Adequate maintenance and process control.

efficient lighting in the residential and service sectors.
- A decrease in the high mountain and forest-steppe area and an increase in the steppe area is expected as a result of climate change. The desert area is projected to increase by 6.9–23.3% of the actual area by 2040 and by 10.7–25.5% by 2070. The findings of GCM scenarios show that water resources will tend to increase in the first quarter of the century and then decrease, returning close to current levels by 2050.
- The impact assessment indicated that temperature increase will have a negative impact on ewe weight gain (1–20 g/day) in all geographical regions because the hot temperature during the daytime will cause a reduction of grazing time.
- The potential wheat yield is expected to increase by 8–58% by 2040 and the potato yield by about 2–26%.
- The area of permafrost will be decreased significantly if warming trends continue. Accordingly, significant changes will take place in the surface water balance, the soil moisture and temperature regimes and the vegetation cover.
- Important adaptation measures include: development of rangeland and livestock management systems; improvement of forage production systems; use of modern pasture water supply systems; establishment of a risk management system; strengthening of the early warning system within the National Meteorological and Hydrological Services; irrigation; improvement of land cultivation management systems; set up of legislative mechanisms for pasture use; and development of an insurance system for livestock and crops with respect to natural disasters.

3.7.6 Senegal

The Initial National Communication of Senegal was published in 1997 (République du Sénégal, 1997), reflecting the country's sense of urgency regarding climate issues. In 1999, the National Communication was followed by a national strategy on climate change (République du Sénégal, 1999.) The National Communication, which appeared in French only, includes six chapters: National Circumstances (21 pp.); Greenhouse Gas Emission Inventory (51 pp.); Vulnerability Studies (15 pp.); Mitigation Strategies (14 pp.); The Climate Change Training Programme (2 pp.); and Conclusions (2 pp.). The NCCSAP contributed to the chapter on vulnerability (agriculture and coastal zones).

The main conclusions of Senegal's National Communication are:

- The total GHG emissions amounted to 3.3 million t CO_2 equivalents (0.4 t/capita) in 1994. Land use change resulted in a net uptake of –6.0 million t. Energy production, agriculture and waste contributed 3.8, 3.0 and 2.2 million t CO_2 equivalents, respectively.
- The coastal zones of Senegal are vulnerable to SLR. Main effects include erosion, saltwater intrusion and mangrove degradation. If the sea level in 2100 has increased by 1 m, 6000 km^2 of coastal area are at risk, 109,000–178,000 people must move, US$500–700 million may be lost and coastal protection costs would amount to US$1–2 billion.
- Without climate change, Senegal is able to feed its own population with moderate intensification of its agricultural practices and without endangering its natural resources in 2050. A drier climate, however, would lower agricultural yields by 11–38%.
- Several GHG mitigation options are investigated, including use of natural gas for electricity production and replacing firewood with LPG.

3.7.7 Yemen

Yemen's *Initial National Communication* was published in 2001 (Environmental Protection Council, 2001). It includes six chapters: Introduction (2 pp.); National Circumstances (21 pp.); National Greenhouse Gas Inventory (12 pp.); Observation, Research and Impact of Climate Change (20 pp.); Measures to Reduce GHG Emission and Adaptation to Climate Change (14 pp.); and

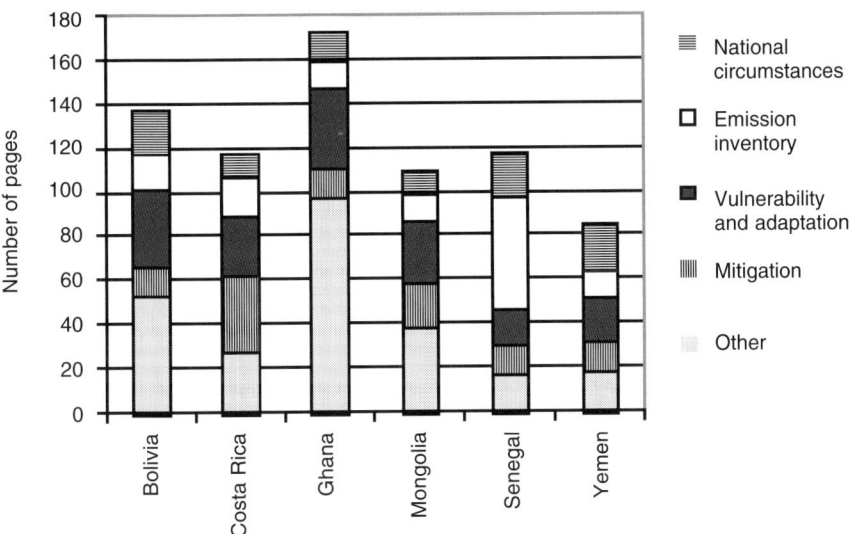

Fig. 3.1. Number of pages in the National Communications devoted to the indicated chapters.

Climate Change Research Needs (4 pp.). The NCCSAP contributed to the vulnerability and adaptation sections regarding agriculture and water resources.

The main conclusions of Yemen's National Communication include:

- In 1995, Yemen emitted 18.7 million t of CO_2 equivalents (1994 data were not available because of the civil war). Main contributors were the energy sector (10.1 million t) and agriculture (6.3 million t). Because 10.5 million t of CO_2 were sequestered in 1995, the net emission amounted to 8.2 million t of CO_2 equivalents.
- Climate models and demand projections indicate a large deficit in the coverage of spate irrigation water. The demand varies between 551 and 595 million m³/year, while the available flows are only 105–127 million m³/year. Therefore, a sharp increase in groundwater use is foreseen, resulting in rapid depletion and salinization of aquifers.
- All climate models project a decrease of potato and wheat yield. The decreases amount to 6–33%, depending on the region. An additional problem is that irrigation needs to be intensified.
- In the coastal area (Hodeidah city) potential losses amount to US$1.3 billion.

- Mitigation options include high efficiency combined-cycle gas turbines, solar water heating, LPG stoves in the place of fuel wood stoves and switching to natural gas and solar technologies.
- Adaptation options include improvement in water use efficiency, groundwater recharge and desalinization; changing the crop yield subsidies, sustainable pest prevention programmes; and coastal protection measures for Hodeidah city.

3.7.8 Synthesis

The National Communications of the countries considered here reflect their diversity. Availability of financial support, international assistance, data and national commitment determined the absolute and relative amount of information provided by the National Communications (Fig. 3.1). Since the guidelines for National Communications were only specific about the emission inventories, this chapter is the easiest to compare. Only Senegal provided much more information than required.

Costa Rica, Ghana and Yemen used their National Communications to seek financial support by including a list of climate projects. On average, 27 pages (22%) were spent on

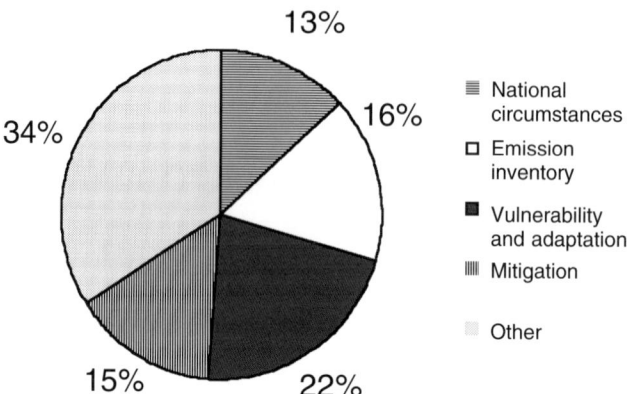

Fig. 3.2. Average share of the indicated chapters in the National Communications.

vulnerability and adaptation, while 18 (15%) pages were spent on mitigation (Fig. 3.2). This suggests that the countries involved have used the National Communication to put the issues of mitigation and adaptation on the international agenda, and that they consider vulnerability and adaptation more important than mitigation. Considering the magnitude of the envisaged impacts of climate change in most developing countries, it must be concluded that in the subsequent version of the National Communications even more attention should be given to the vulnerability and adaptation sections.

4
Evaluation, Lessons Learned and Outlook

4.1 Introduction

This chapter aims to put the results of the Netherlands Climate Change Studies Assistance Programme (NCCSAP) and pre-NCCSAP studies into a wider perspective: how developing countries can effectively deal with climate change in cooperation with developed countries or donor programmes. It discusses the methods used in the studies, provides future outlooks and examines the issues of awareness, capacity building and policy implementation.

The sections in this chapter are Mitigation Assessment (Section 4.2), Adaptation Assessments (Section 4.3), National Communications (Section 4.4), Capacity Building and Awareness Raising (Section 4.5) and NCCSAP in Comparison with Other Country Study Programmes (Section 4.6). The chapter ends with Recommendations and Conclusions (Section 4.7).

4.2 Mitigation Assessment

Nico van der Linden[1] and Jan-Willem Martens[2]

4.2.1 Introduction

As part of the NCCSAP, mitigation studies have been implemented with the objective of promoting the further implementation of mitigation activities in non-Annex I countries. As described in Sections 1.4 and 3.3, such studies have been implemented in Bolivia, Mongolia, Yemen and Zimbabwe. The purpose of this section is to put these efforts into a wider perspective of greenhouse gas (GHG) mitigation activities in developing countries.

The conclusions from these four countries are quite illustrative of the experience with mitigation policies in developing countries. While developing countries' emissions continue to grow, a switch to more GHG neutral technologies is not yet foreseen. It is highly unlikely that many climate friendly technologies would be developed in many developing countries due to unfavourable market conditions and the existence of significant market barriers. This was recognized in the Intergovernmental Panel on Climate Change (IPCC) Special Report on Technology Transfer, which mentions the slow progress in the development of many climate friendly technologies in non-Annex I countries as a result of a number of policy, technical, financial, management, institutional and awareness barriers. For example, the key barriers for the development of renewable energy include a lack of technical expertise and weak institutional structures to plan, manage and maintain renewable energy programmes; the absence of clear policies and plans to guide renewable energy development;

---
[1] See Note 2, Chapter 1.
[2] See Note 3, Chapter 1.

a lack of successful demonstration projects; a lack of potential renewable energy resources; a lack of confidence in the technology by the policymakers; a lack of local financial commitment or support for renewable energy; and continuing reliance on aid-funded projects. In fact, most initiatives regarding the development of renewable energy projects are either supported by ODA programmes or by the Global Environment Facility (GEF). The GEF, through its implementing agencies (the World Bank, the United Nations Development Programme (UNDP), the United Nations Environment Programme (UNEP) and the United Nations Industrial Development Organization (UNIDO)) has initiated a large number of projects and programmes in an attempt to reduce implementation costs and barriers to project development of renewable energy. This further demonstrates that commercialization of climate friendly technologies in non-Annex I countries will not occur without external assistance and the provision of financial incentives.

This process is confirmed if one looks to projections of the adoption of GHG neutral technologies in developing countries. The growth projections for the power sector (one of the largest GHG emitting sectors) are an example of this. Figure 4.1 shows the projections of energy sources used to generate power up to 2030. It is clear from this figure that the low GHG emitting technologies (nuclear, hydro and other renewables) have a constant share of around 15% in the total electricity generation in developing countries. The majority of this 15% are provided by nuclear and hydro, which are associated with other environmental problems. Note that the largest contribution to lower GHG emissions is expected from a switch from oil to natural gas.

Despite the bleak outlook for mitigation activities, there are a number of autonomous trends that will result in the relative reduction of GHG emissions.

- Energy efficiency measures across a wide diversity of industries. Market parties will have an incentive to focus on energy efficiency, especially in those countries where subsidies for energy are removed.
- Fuel switch from oil and coal to natural gas in the power sector and transport (as is demonstrated in Fig. 4.1).
- Niche applications for renewable energy that can compete with conventional energy sources are:

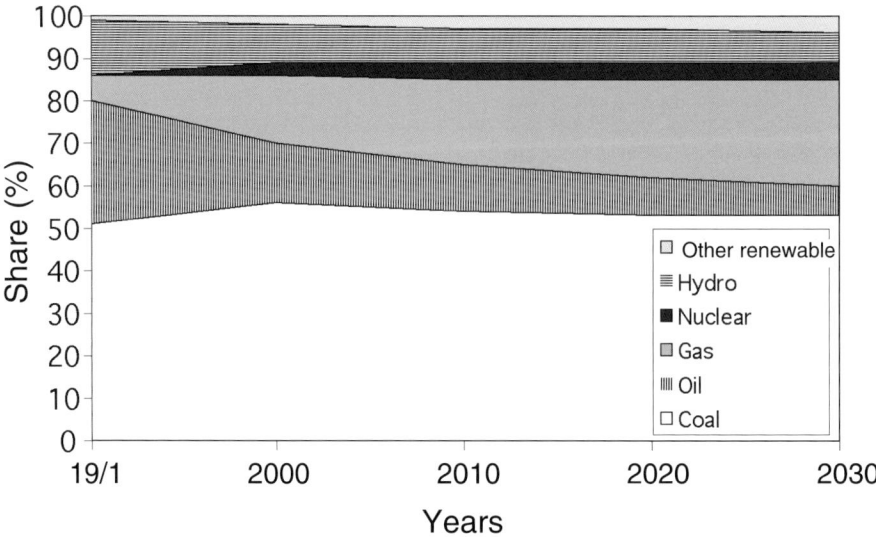

Fig. 4.1. Energy sources for power generation for the next 30 years in non-Annex 1 countries (Source: World Energy Investment Outlook, 2003).

- the generation power of hydropower plants, both on a large and small scale;
- rural mini-electricity grids powered by wind, solar, solar-wind hybrids or wind-diesel hybrids;
- household energy technologies for rural households such as home solar systems, improved cooking stoves;
- solar water heaters for urban households which have been quite popular in countries such as Botswana and Nepal;
- stand-alone power for remote telecommunications infrastructure;
- cogeneration in sugar mills using bagasse residues;
- wind energy in countries with supportive local policies, such as in India.

Nevertheless, despite such positive trends, a majority of important GHG emissions in non-Annex I countries will not be addressed. This is a concern because developing countries' emissions continue to grow and there is too little stimulus to create a structural shift towards a low carbon economy. Indirect policies are less effective than direct policies, and GHG emission reduction is often not the main target of policies. Non-hydro renewable energy power projects, CH_4 reduction of wastewater treatment and landfill gas extraction, and reduction of industrial GHG such as N_2O and HFC-23 are among the cost-effective solutions to decrease GHG emissions in non-Annex I countries.

4.2.2 Recent developments with regard to the Clean Development Mechanism (CDM)

Current status of CDM

As of the middle of 2004, no CDM projects had passed the entire approval process for registration as official CDM projects. However, Project Design Documents (PDDs) have been prepared for a number of projects; these projects are denoted as 'CDM projects' in this section. Figure 4.2 shows an overview of the current mix of CDM projects.

Figure 4.2 clearly demonstrates that most CDM projects are renewable energy projects. At present, the share of renewable energy, including 'bio/landfill gas', is 73%. Hydropower is the most dominant category of CDM projects. Most projects reduce CO_2 emissions, except the projects in the bio/landfill gas category (16%), which reduce CH_4 emissions, and one project in the industrial sector, which reduces HFC-gases. Regarding the renewable energy mix, it is apparent that there are no solar energy projects and only a few geothermal energy projects (5%). In terms of the share of emission reductions, landfill projects are very popular. In landfill projects CH_4 emissions are reduced, and because 1 kg of CH_4 equals 21 kg of CO_2 in terms of global warming potential, they are relatively cost effective. Furthermore, if the CH_4 produced is incinerated to produce electricity or heat, CO_2 emissions from fossil fuels are prevented.

An interesting aspect is that contrary to expectations, there are relatively few energy efficiency projects. One reason could be that energy efficiency projects are already financially feasible and do not need to be developed as CDM projects. Another explanation is that due to specific barriers for energy efficiency, the required internal rate of return for energy efficiency needs to be much higher than for power supply projects before implementation is considered.

Location of mitigation activities

A second interesting characteristic of CDM projects is the project location. The success of

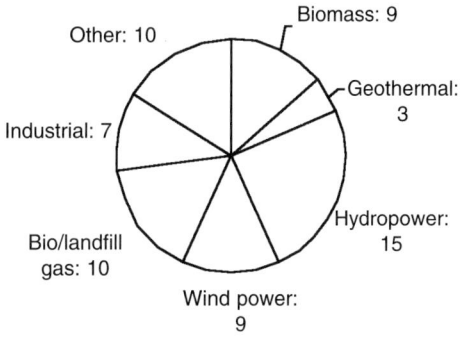

Fig. 4.2. Number of CDM projects per technology category (not the total *emission reductions* of the projects) (Source: EcoSecurities CDM project database).

a CDM project can be greatly influenced by the country in which it is located. Important country-specific CDM factors are:

- the intention of the government to ratify the Kyoto Protocol;
- institutional organization and willingness to endorse the transfer of Certified Emission Reductions (CERs);
- in the case of electricity generation, the average amount of GHG emissions of the existing electricity park, which influences the amount of emission reductions a CDM project can generate.

Apart from these specific factors, factors influencing the broader investment climate for mitigation activities are crucial. According to the IPCC Special Report on technology transfer (Metz et al., 2000), the 'enabling environment' provided by the host country is a very important determinant as a success factor for the implementation of mitigation activities. Elements of what constitutes an enabling environment are the investment climate, such as financial infrastructure and macroeconomic policies, but also soft skills such as technical capabilities of the labour force and the presence of R&D and other innovation institutions in a country.

Conclusion

Even though the current evidence is not substantial enough to have any scientific meaning, it appears that CDM supports the implementation of mitigation activities that otherwise would not be likely to take off. Private sector actors have responded to opportunities to benefit from the CDM, especially in the renewable energy sector and the waste management sector (landfill gas recovery and wastewater treatment). Given the early stage of CDM and the lack of clarity in CDM procedures on the international policy level, this seems quite promising for the financing of mitigation activities in developing countries. As is to be expected, mitigation activities are taking place in those countries that provide the most supportive environment for such activities.

4.2.3 Future developments with regard to CDM

The CDM is expected to become the most important financial source for climate mitigation activities in developing countries in the near future. The most important buyers of CDM credits (or Certified Emission Reductions – CERs) will be EU member states, thanks to the emerging EU Emission Trading Scheme. As mentioned above, the demand for CERs comes from Annex I governments. On the other hand, the EU industries with an emission cap under the EU Emission Trading Scheme will also act as buyers of CERs. Recognizing there are still many factors that may hamper demand for CERs, Table 4.1 provides an estimate of possible demand.

Table 4.1. Demand for CERs up to 2012 (Sources: Private communications with governments; Point Carbon, 2004; Bygrave and Bosi, 2004).

Governments	CO_2 (million t)	
Austria		70 million
Canada		100 million
Belgium		10 million
Finland		10 million
Italy	11	55 million
Spain	130[a]	650 million
The Netherlands	100	500 million
Japan Carbon Fund		46 million
Subtotal		**1.4 billion**
Industries in the EU Emission Trading Scheme		2–8 billion

[a]This figure was indicated by the previous Spanish government lead by the Popular Party. It is not clear if the new Spanish government led by the Socialist Party will continue that policy.

4.2.4 Long-term perspectives on mitigation activities

One of the key lessons of the previous subsections is that the flexible mechanisms of the Kyoto Protocol are the key drivers for the implementation of mitigation activities in developing countries. The combined carrot and stick elements in the Kyoto Protocol (binding commitments for Annex I Parties and flexibility to sell carbon credits for projects in developing countries) provide the incentives for market players to engage in mitigation activities at their own risk. Despite the uncertainty of ratification of Kyoto in the past and the long bureaucratic process to get CDM started, project developers have enthusiastically engaged in developing mitigation projects in non-Annex I countries with the aim of selling carbon credits.

It is important to recognize that there are many steps to be taken before international carbon trading can provide the necessary drive for a structural reduction in global GHG emissions. The most important step for the international community to take is to reduce the current policy uncertainty. At the moment, there is still a highly uncertain market environment with many political risks unknown to other commodity markets. The uncertainty will ensure that major investments in GHG reducing technologies and large-scale dedication of R&D are still too risky. Mainstream market parties have not engaged themselves actively in the market (apart from a few global leaders); most follow the political developments in the carbon market from the sideline. The political uncertainty can be reduced to two elements. First, it is not clear if there will be a long-term international climate policy rewarding structural reduction in GHG emissions. Second, even if a long-term carbon policy is adopted and a long-term carbon trading framework is guaranteed, it is not clear if under such a policy 'early action' to reduce GHG emissions will be rewarded.[3] Until such fundamental policy uncertainties are clarified, investing in GHG mitigation could be an investment with no return. Thus, major market players will likely consider a 'wait-and-see' approach as their best strategy.

The second Kyoto commitment period

In the first Kyoto commitment period (2008–2012), the historic emissions are the starting point for calculating the emissions' ceilings. Countries have a reduction target relative to their 1990 emissions. The actual reduction targets are different per country and depend on domestic reduction opportunities and negotiating positions in Kyoto. The Kyoto ceilings range from −8% for the EU (as a whole) to +10% for Iceland. Developing countries do not have any targets.

Unfortunately, future commitments for a further mitigation of GHG emissions after Kyoto are still in a preliminary stage in the international climate change arena. In renegotiating ceilings, an important issue will be whether developing countries will take up targets as well. Strict reduction targets that would harm economic growth will not be acceptable to developing countries and would not be fair considering the much higher per capita emissions of some developed countries. However, one reason the US and Australian governments gave for not ratifying Kyoto was the fact that developing countries (more specifically China, India and Brazil) have no targets at all. If developing countries take up targets (even targets allowing for growth of GHG emissions), this would help to reach a political agreement. Some developing countries are already considering this. For example, in its domestic climate policy, Malaysia is considering the fact that it will have to assume a ceiling under the second commitment period. Also, Chile and Argentina have indicated in the past that they are favourable to the concept of adopting binding commitments. So far, serious progress on negotiating a second commitment period has not been

---

[3] To illustrate this point, take the EU Emission Trading Scheme: most national allocation plans for the first period (2005–2007) are based on allocating historic emissions for free (a principle called grandfathering). Since inefficient industries have relatively more historic emissions compared to their more efficient competitors, they have been allocated more emission allowances. The efficient industries have thus been penalized for cutting their CO_2 emissions prior to the Emission Trading Scheme coming into force.

made, even though the Kyoto Protocol has come into force.

A framework for long-term climate policy: contraction and convergence

Regardless of the progress in international commitments, there are a number of approaches in the literature providing a long-term framework for international GHG emissions' trading, aimed at a long-term reduction of GHGs. Contraction and Convergence (C&C) is an idea that is promoted by the Global Commons Institute. The aims are to avoid global climate destabilization and to do this in an equitable way.

The first part, *contraction*, starts with the assumption that there is a certain safe level of GHGs in the atmosphere. If this level is exceeded, the world would risk catastrophic effects of climate change. It is difficult to say exactly what the safe level is, but it is commonly agreed that CO_2 concentrations in the atmosphere should stay within the range of 450–550 parts per million by volume (ppmv). On the basis of this, the maximum level of worldwide CO_2 emissions over time can be calculated.

To be realistic, contraction should take into account the current CO_2 emissions and also the growth path of emissions in the short term. In the longer term, there has to be a large contraction of emissions in order to stay within the safe level of (for example) 450 ppmv in the atmosphere. Based on the agreed upper limit of CO_2 concentration combined with a feasible rate of emission reductions over time, a global emissions budget can be set.

The second part, *convergence*, is about an equitable distribution of the worldwide emissions budget. The ideal would be an equal per capita distribution of emission entitlements. This could be done per year, and distributed per country. The emission entitlements should then be tradable between individuals and/or between countries. Given population growth and the fact that emissions have to be reduced over time, the per capita entitlements will become less each year.

A sudden introduction of an equal per capita distribution of emission entitlements would not be politically acceptable. The current per capita emissions in developed countries are many times higher (especially in the USA) than those in developing countries. A trading system combined with a limit based on equal per capita emissions would involve huge transfers of money to developing countries.

Implication for global mitigation activities

An international Emission Trading Framework based on or similar to the Contraction and Convergence Framework has a number of benefits as a basis for international climate policy:

- By assigning property rights to all global GHG emissions and by providing binding commitments for all countries, it provides the backbone for a global long-term carbon market.
- Emission rights could be allocated at a national level, based on an equal per capita distribution worldwide, and any activity reducing emissions would have a direct value. Hence, it would not matter if emissions are reduced via some national policy or a project. Any developing country reducing emissions in any way would be rewarded, because it could sell its allowances on the global carbon market.
- It provides market players with a long-term framework for assessing the development of GHG policy on the basis of which they can make long-term investment decisions. The long-term nature of the policy framework will send a clear price signal to market players, allowing them to invest more in R&D of GHG reducing measures and in the application of more rigorous breakthrough technologies.
- A distribution of emission rights based on per capita emissions would result in developing countries receiving excess allowances that they could export to industrialized countries that are in need of additional allowances. This long-term transfer of cash could provide the financing for long-term investments in CO_2-neutral investments, thus promoting a cleaner development path.
- A per capita distribution of emission rights at the national level would be equitable and

also solve the problem of perverse incentives that currently exists under the CDM. The current interpretation of *additionality* is a disincentive for developing countries to develop decarbonizing policies.

In CDM the concept of additionality means that a project is only considered additional if it reduces emissions compared to the situation without the CDM project, the baseline. This baseline also includes the legal requirements a country imposes. If climate-friendly sectoral policies and laws are put in place, projects are considered non-additional and are therefore excluded from the CDM. This provides non-Annex I governments with a perverse incentive to refrain from introducing climate friendly legislation, in order to keep attracting investments via CDM.

With regard to mitigation activities, Fig. 4.3 gives a picture of how the targets could be reached while continuing business-as-usual economic growth. The thick line shows the business-as-usual emissions. By increasing efficiency the existing emissions growth trend could be reduced. The largest part of emission reductions will have to come from renewables. The graph also shows that in the process of adapting to low emission levels, emission allocations can be traded between developed and developing countries.

4.3 Adaptation Assessments

Marcel Rozemeijer[4], Arjan van der Weck[5], Frank van der Meulen[6], Rob Misdorp[7], Hans Leenen[8], Jan Verhagen[9], Frits Mohren[10] and Michiel van Drunen[11]

4.3.1 Coastal zones

As shown in Section 3.5, the country assessments gave insight into the degree of vulnerability of the country's coastal zones. Feasibility estimates indicated the degree to which countries are able to develop response strategies to make them less vulnerable in the future. This section briefly discusses some general aspects that need to be taken into consideration when looking at future adaptation strategies. The aspects are interconnected.

Awareness raising and stakeholder participation

So far, most results are only known to the experts who worked in the studies. To gain

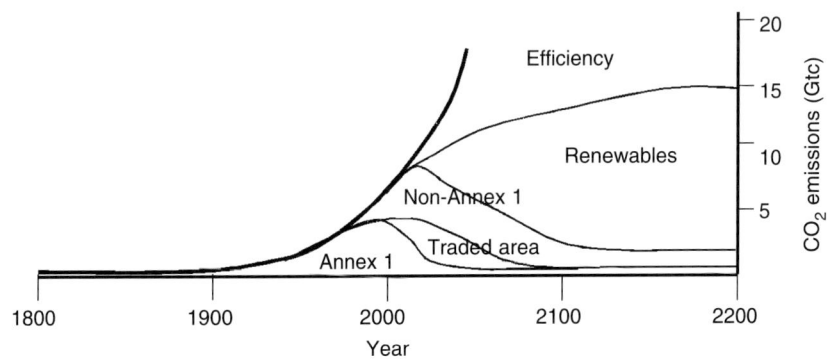

Fig. 4.3. Share of renewables in the business-as-usual scenario (Source: http://www.gci.org.uk/).

[4] See Note 8, Chapter 1.
[5] See Note 9, Chapter 1.
[6] See Note 9, Chapter 3.
[7] See Note 10, Chapter 3.
[8] See Note 5, Chapter 3.
[9] See Note 5, Chapter 1.
[10] See Note 7, Chapter 1.
[11] See Note 1, Chapter 1.

political support for planning and implementation of response strategies, a much wider stakeholder group needs to be involved. Two strategies to gain political support are:

- linking adaptation measures for sea level rise (SLR) to current economic and social interests;
- linking the impacts of climate change to the daily livelihood of coastal communities and enabling local stakeholders to claim ownership of the project results.

When adaptation measures have to be taken, it is important to identify the stakeholders that are associated with such activities. Who are they? Do they see the measure as solving a particular problem? Are they committed to cooperate in the action with other stakeholders? Such 'ownership' is vital for effective planning and implementation of measures. In most cases, a change in behaviour of the stakeholders is desirable for effective coastal management. An example of a method to achieve such behavioural changes can be found in Olsen (2003).

Spatial planning

Planners are important stakeholders. In a larger perspective, effective coastal zone management is a matter of spatial planning. Hence, questions that must be addressed in assessments are: What are the planning scenarios for those areas that are at risk? If the use of natural dynamics of the coast is an option in dealing with the consequences of SLR, what are the plans for building hard obstacles and structures in the coastal zone? Impacts have been studied for periods up to 30–100 years. Furthermore, it is important that the planning focuses on no-regret solutions rather than on options for short-term gains (IPCC, 2001).

Case studies and 'work with nature'

Commitment and awareness are raised when communities see that measures are being planned and implemented that really improve their quality of life in the sense of safety, natural resources and socio-economy (Olsen, 2003). (Small) case studies could offer experience in this. It is proven good practice in coastal management that such projects develop best at community or small provincial levels, where the needs are/will be felt most strongly. Such projects need the support of higher governmental organizations (a vertical integration of decision layers), preferably under a national Integrated Coastal Zone Management (ICZM) strategy. In low-lying, vulnerable coastal areas there is a need for more detailed vulnerability assessments and response strategies involving not only traditional hard engineering options, but also 'soft' engineering options (e.g. sand nourishment). Making use of the natural dynamics of the coast ('working with nature') could present interesting response options. This has consequences for spatial planning.

Integrated Coastal Zone Management

The present vulnerability of the coast will be increased by the impacts of climate change and SLR. ICZM is seen as the most suitable approach to cope with the complexity of problems along the coast. Adaptation should be embedded in ICZM. It covers technical, political and institutional aspects. The establishment of a high level coordinating ICZM institution will favour this process. Box 4.1 provides an example of ICZM as an adaptive strategy.

Outlook

The IPCC methodology has been used frequently, not only within the NCCSAP, but also in many other studies. It offers a generic framework comprising seven main steps of analysis, designed to be applicable to any natural and socio-economic system potentially affected by climate change. Klein et al. (1999) published a critique that focused specifically on Step 6 (assessment of autonomous adjustments) and Step 7 (evaluation of adaptation strategies). They describe experiences in the Netherlands, the UK and Japan, which more or less coincide with the experiences described above. They note that important missing elements include:

- the interaction between climate change and other pressures in determining impact potential;

Box 4.1. Vietnam as an example: a vulnerable coastal nation, practising vulnerability assessment and ICZM programmes as adaptive strategies.

The coastal zone of Vietnam is critically vulnerable to SLR, while the financial coping capacity is relatively limited (Section 2.12). Among the 179 assessed coastal countries Vietnam ranked in the top class of coastal vulnerabilities. This was confirmed by the *Vietnam Coastal Zone Vulnerability Assessment* 1994–1997 (Hydrometeorological Services, 1996), which concluded that ICZM was recognized by the government of Vietnam as an appropriate, adaptive strategy enhancing sustainable development of coastal resources. Between 1998 and 2000 preparations were made for a follow-up ICZM study in Vietnam. The CZM-Centre/RIKZ (Netherlands Ministry of Transport, Public Works and Water Management), along with representatives of the Vietnamese Ministry of Environment and the Netherlands Ministry of Foreign Affairs, prepared the Terms of Reference for an ICZM project in line with recommendations of the United Nations Conference on Environment and Development (UNCED) – AGENDA 21 and the 1992 Framework Convention on Climate Change. The main Vietnam Netherlands (VN)-ICZM project encompasses the period from 2000–2003. Vietnamese coordination is in the hands of the Ministry of Science, Technology and Environment and, since 2003, the Vietnamese Ministry of Natural Resources and Environment (MONRE). The ICZM expertise input is funded and coordinated by the Royal Netherlands Embassy in Hanoi and directed to Strategy and Action plans, data collection and management, and awareness raising. The supporting Coastal Cooperation Programme (CCP) is a cooperative effort of two ministries: the Netherlands Ministry of Transport, Public Works and Water Management and the Vietnamese MONRE. It provides assistance in VN-ICZM institutional building at the national level and supports VN-ICZM awareness raising, monitoring, training and sustainable restoration at TTHue provincial level. At present, preparations are being made for the second phase of VN-ICZM and CCP, starting in 2005. Results of these VN-ICZM efforts include:

1. Institutional strengthening: the simultaneous execution of the ICZM programmes at the national level and in three coastal provinces resulted in a clear increase of vertical integration through institutional strengthening, essential for an ICZM programme. The high level national ICZM Committee is chaired by the Vice Minister of MONRE and consists of Vice Governors (Vice Chairmen of the Provincial People's Committee) of the three ICZM pilot provinces and Directorate Generals of participating relevant ministries. The legal mandate of this high level ICZM Committee is expanding, the input of NGOs and local stakeholders is being debated, and the coordination of international funding for the implementation of ICZM based projects has been strengthened. An ICZM Division recently created within MONRE will ensure assistance in the planning and implementation of ICZM programmes in future.
2. Horizontal integration has taken shape through intensified cooperation and coordination between the relevant provincial departments and universities, local and international experts. The provincial ICZM Strategy Plans and Action Plans have been prepared in a consultative way, using the integrated expertise.
3. Awareness raising: VN-ICZM newsletters are distributed among all Vietnamese coastal provinces. A website linked to www.netcoast.nl provides an overview of activities and coastal databases. Several training courses on restoration and management of wetlands, on ICZM, and on tools such as Remote Sensing and GIS were organized. Drawing competitions among 450 schoolchildren were held in TTHue province. The 20 best drawings served as illustrations for the teachers: 'Where waters and land meet'. A series of ICZM consultative platform meetings at district or communal level sparked discussions on local problems and solutions among a wide audience mixing local, provincial and international expertise.
4. Monitoring programme: in TTHue province a monitoring programme was started on coastal dynamics, environmental quality and biodiversity, with a strategic policy document focusing on 'Why to monitor'. It included the expectations of the Provincial and District Leaders about dealing with overexploitation, pollution and coastal erosion. The results of the first systematic monitoring campaigns show that an increase in knowledge about coastal processes is beneficial for decision making on the sustainable development of near shore and lagoon resources.
5. From river to coast: some impacts of processes and interventions in the mountainous hinterland and the coastal zone in TTHue province were illustrated by STREAM, a GIS-based water balance model. The

> output of STREAM, including impacts of climate change, was used as input for the coastal 3D-lagoon modelling.
>
> In conclusion: 10 years of ICZM efforts in Vietnam have revealed that ICZM is indeed a useful mechanism in addressing both the short-term challenges related to population pressure and the long-term impacts of climate change. The results of the ICZM efforts in Vietnam were extensively discussed during provincial and national ICZM workshops. The results were reported, internationally reviewed and positively evaluated. The simultaneous introduction of ICZM at the national level and in the three pilot provinces was successful. Simultaneous execution of higher level aggregated activities (strategy formulation) and lower level activities such as setting up monitoring programmes, local platform discussions aimed at the preparation of workable projects is tedious but fruitful, and requires leaders who have a good overview of the situation. The introduction of ICZM has required persistence on the part of the ICZM leading agencies, as many conflicting interests need to be accommodated. Careful preparation takes time, several years in fact, as has been clearly demonstrated in Vietnam. The holistic approach of ICZM is appreciated in Vietnam and will further be introduced in the next set of the 29 low-lying coastal provinces.

- public perception and awareness of climate change and its impacts;
- spatial and temporal planning of adaptation measures;
- mechanisms for public involvement;
- non-technical (i.e. economic, legal, institutional) aspects of adaptation; and
- tools and procedures to evaluate adaptation performance.

Furthermore, Klein et al. (1999) regard coastal adaptation to climate change as a multi-stage and iterative process. Four basic steps recur in each of the case studies:

1. Information collection and awareness raising.
2. Planning and design.
3. Implementation.
4. Monitoring and evaluation.

According to Klein et al. (1999):

> Assessing these elements in addition to the steps prescribed in the IPCC Technical Guidelines will give a more complete picture of a country's adaptive capacity and hence of the range of actions required to reduce its vulnerability ... These actions are expected to be more than the implementation of technical options to retreat, accommodate or protect.

As indicated in the coastal zone management studies described in this book, geographic information systems (GIS) provide an essential tool. To support the 'multi-stage and iterative process', GIS and maps can be used as negotiation tools to support water management (e.g. Janssen et al., 2005). The idea is that stakeholder preferences are translated into spatially explicit maps. On the basis of these maps, proposed plans can be evaluated and potential conflicts can be identified.

Such activities can be placed within the broader context of what has been called Participatory GIS (PGIS). Participatory GIS has its roots in the 1990s, when several social theorists started to express concerns about the claimed objectivity and value-neutral nature of GIS (for an overview of the various authors and their arguments see Sheppard et al., 1999 and Weiner et al., 2002).

Specific questions addressed in this body of literature included: how current models of geographical space have been developed in existing GIS; which cognitive and social understandings of geographical space may have been left out; how these lacunae have affected the utility of GIS for different groups or purposes; how these lacunae can be filled; how communities can be more involved; and how the use of GIS affects certain groups or individuals in society, i.e. whether it leads to empowerment or marginalization.

As the PGIS research is in its infancy, it is, at this moment, still difficult to draw firm conclusions about the usefulness of PGIS as an approach to climate change vulnerability assessment. Nevertheless, it can be concluded that PGIS holds great potential and it is very likely that it can stimulate the public debate and increase the involvement of people in the decision-making process.

4.3.2 Water resources

Evaluation

As indicated in Section 3.4, the experiences with water resources varied from country to country. In all countries, specific case study areas or specific river basins were assessed. Therefore, no complete picture for the countries as a whole could be established. Generally the models were sufficient for the level of detail required. Also, the data requirements were not too high.

Water resources are crucial for many sectors including agriculture, industry and electricity production. Therefore, an integrated, multidisciplinary approach is crucial for a comprehensive analysis and the formulation of effective adaptation options. In some of the studies, the cooperation with experts from different sectors appeared to be problematic.

A problem that was encountered in some of the water resources studies was a lack of training in the programme, as some of the studies were not foreseen in the original project proposals. The training was given to technical committees only on an *ad hoc* basis. The need to make room for additional training in the project plans was expressed during the Amsterdam workshop (Dorland *et al.*, 2001). This training should be multidisciplinary and open to specialists and non-specialists.

Outlook: water pricing

The Dublin Statement of the International Conference on Water and the Environment (ICWE, 1992) states 'water has an economic value in all its competing uses and should be recognised as an economic good'. But internationally there is little agreement on what this actually means, either in theory or in practice. The interaction of three critical factors – the value of water, the use cost of water and the opportunity costs of the resource – should be explored and assessed for different situations. The assessment of the relative magnitudes of 'use costs' and 'opportunity costs' shows that the implications of treating water as an economic resource vary widely depending on the sector.

Water scarcity not only arises from lack of available resources, but also through uncontrolled and unsustainable use, or water spillage. Potential adaptation measures in this case could be directed to control the use of water. The important challenges for urban water utilities in developing countries are, therefore:

- to reduce costs by more efficient operations (which increasingly means substantial involvement of the private sector);
- to raise tariffs (which typically cover less than one-third of costs).

A next step in the development of adaptation strategies for water resource problems could be to focus on the key issues of water scarcity. The potential for improvement of existing situations and combating the anticipated worsening of those situations have not been exhaustively analysed. Water scarcity is a common problem and is not just associated with climate change effects. It is already very recognizable, whereas the risk of flooding in areas that are not suffering from floods at the moment is a rather abstract concept. Thus, a next step could be to investigate the potential for water demand management by recognizing water as an economic good. This is a rather sensitive but challenging concept and in some form or another it is worth opening it up as a valuable discussion point for potential adaptation strategies.

4.3.3 Agriculture and forestry

Climate change and climate variability can substantially affect agricultural production and may have a dramatic impact on local and national economies. Vulnerability of the agricultural sector to environmental change is, however, a complex issue. When addressing issues of food supply and food security, production levels alone are not the only issue. Socio-economic mechanisms need to be addressed as well. Most studies focused on the direct impact of climate change on crop production at the farm level. Inclusion of possible effects on farm management and socio-economic and socio-cultural structures were not fully considered (Dorland et al., 2001).

Isolating the effects of climate change on complex systems is not a trivial matter. The IVM/UNEP handbook (Feenstra et al., 1998) provided to the teams emphasized the importance of an integrated assessment. It did not, however, provide the tools to perform this complex task. A stronger farming systems component linked to socio-economic studies would improve the impact and adaptation studies.

The NCCSAP clearly contributed to strengthening cooperation between different national institutes, and it also established a south-south network. Two other objectives, bringing together expertise from various disciplines and producing an integrated assessment, were not always achieved. Linking technical and socio-economic studies requires a common integrated framework.

Outlook

Future work may focus on the evaluation of forest and agricultural land use policies and their effect on the vulnerability and emission profiles of land use systems. Formulating development pathways while taking changes in climate and climate variability into account is a way to have a more integrated approach to the climate change issue. Also, in such an approach results from simulation models or semi-quantitative models can be used to set priorities and reveal trade-offs between response strategies. The framework used so far does not stimulate radical changes in the design and operation of land use systems.

Issues such as bio-products, bio-fuels and carbon sequestration are currently not included in the agricultural studies. These issues may emerge when moving from adaptation to mitigation. This also holds for energy and water use efficiency in the agricultural production chain (Dorland et al., 2001).

4.4 National Communications

Michiel van Drunen[12]

4.4.1 Guidelines

As indicated in Subsection 3.7.1, the 'guidelines' for writing National Communications for non-Annex I countries were very brief at the time the NCCSAP-I studies were conducted. The new guidelines that were established in 2003 (UNFCCC, 2004a) not only describe the reporting requirements of emission inventories (9 pages), but also explicitly deal with adaptation (8 pages) and mitigation (5 pages). Since the methodologies of emission inventories and mitigation are relatively straightforward and well established, the focus will be on methodologies for vulnerability and adaptation. The initial National Communications of the NCCSAP countries showed a wide variety of approaches and level of detail in the sections about vulnerability and adaptation (see Subsection 3.7.8), partly because the assessment methodologies needed to be developed in parallel to the assessments themselves.

In the adaptation section of the new guidelines, the IVM/UNEP handbook (Feenstra et al., 1998) and the IPCC Seven Steps Method (Carter et al., 1994) are explicitly mentioned in a footnote as examples of 'appropriate methods and guidelines ... for assessing vulnerability and adaptation to climate change'. The handbook and the Seven Steps Method were discussed in detail in Subsections 1.5.2 and 1.5.3. The new guidelines also refer to the UNDP Adaptation Policy Framework (APF) and the National Adaptation Programmes of Action (NAPA) guidelines. These two sets of documents are summarized in the *Compendium on Methods and Tools to Evaluate Impacts of Vulnerability and Adaptation to Climate Change*, which is published on the United Nations Framework Convention on Climate Change (UNFCCC) website (UNFCCC, 2004b).

---

[12] See Note 1, Chapter 1.

4.4.2 The UNDP Adaptation Policy Framework (APF)

The APF aims:

> to provide guidance to developing countries for formulating national policy options for adaptation to climate change. A major focus of the APF is to help countries integrate such adaptation policies into national and sectoral planning. It describes the key analytical concepts for developing adaptation strategies, policies and measures. The framework was initiated by UNDP in response to developing countries needs and builds upon the vulnerability and adaptation assessments conducted within the initial National Communications of non-Annex I Parties.
>
> (UNDP, 2004)

The APF provides guidance on designing and implementing projects that reduce vulnerability to climate change, by both reducing potential negative impacts and enhancing any beneficial consequences of a changing climate. It seeks to integrate national policy-making efforts with a 'bottom-up' movement. The framework emphasizes five major principles:

- Adaptation policy and measures are assessed in a developmental context.
- Adaptation to short-term climate variability and extreme events are explicitly included as a step towards reducing vulnerability to long-term change.
- Adaptation occurs at different levels in society, including the local level.
- The adaptation strategy and the process by which it is implemented are equally important.
- Building adaptive capacity to cope with current climate is one way of preparing society to better cope with future climate.

The APF is a flexible approach in which the following five steps may be used in different combinations according to the amount of available information and the point of entry to the project (UNDP, 2004):

1. Defining project scope and design.
2. Assessing vulnerability under current climate.
3. Characterizing future climate related risks.
4. Developing an adaptation strategy.
5. Continuing the adaptation process.

The framework focuses on the involvement of stakeholders at all stages. The key tools include vulnerability mapping, dynamic simulation of sustainable livelihoods, multi-stakeholder analysis, cost–benefit analysis, decision trees and multicriteria analysis (UNFCCC, 2004b).

4.4.3 The National Adaptation Programmes of Action (APA) guidelines

NAPAs are mandated by the Conference of Parties to the UNFCCC for countries participating in the United Nations Least Developed Countries (LDC) Fund. This fund provides financial resources for the NAPAs. Countries are required to rank adaptation measures for funding based on such criteria as urgency and cost effectiveness. The NAPA guidelines are not in themselves a detailed framework for the assessment of vulnerability and adaptation. Instead, they provide some guidance for the process of compiling a document that specifies priority adaptation actions in the least developed countries (UNFCCC, 2004b).

The rationale behind the guidelines is that currently it is not possible to accurately predict climate change and its adverse effects, particularly at the local and regional levels. The IPCC maintains very strongly that learning to deal with climate *variability* and *extremes* is an excellent way of building adaptive capacity in the long run.[13] Strategies to cope with current climate variability and extremes exist at the community level. Hence one of the functions of the NAPA is to identify urgent action needed to expand the current coping range and enhance resilience in a way that would promote the capacity to adapt to current climate variability and extremes, and consequently to future climate change.

The NAPA guidelines explicitly advocate the participation of men and women at the grassroots level: 'First, they are able to provide information on current coping strategies that the NAPA seeks to enhance. Second, they will

---

[13] This was also explicitly acknowledged by Hernán Moreano (Subsection 2.4.6).

be affected the most by climatic impacts and hence will benefit the most from the actions prioritized in the NAPA' (Least Developed Countries Expert Group, 2002).

The guiding elements imply that the NAPA process should emphasize:

- a participatory approach involving stakeholders;
- a multidisciplinary approach;
- a complementary approach that builds on existing plans and programmes;
- sustainable development;
- gender equity;
- a country driven approach;
- sound environmental management;
- cost effectiveness;
- simplicity; and
- flexibility based on country specific circumstances.

In the NAPA process, much of the work of assessing vulnerability and adaptation is intended to be drawn from existing sources. The guidelines do stress the importance of conducting a participatory assessment of vulnerability to current climate variability and extreme events as a starting point for assessing increased risk due to climate change. The key tools used in NAPAs are the same as in the APF (UNFCCC, 2004a).

4.4.4 Discussion

Whereas the IPCC technical guidelines (Carter et al., 1994) and – to a lesser extent – the IVM/UNEP handbook (Feenstra et al., 1998) represent examples of first-generation approaches to the assessment of vulnerability and adaptation, the APF and the NAPA approaches are examples of second-generation assessments that place the assessment of vulnerability at the centre of the process. While the first-generation guidelines have an analytical thrust, and focus on an approach that emphasizes the identification and quantification of impacts, the second-generation guidelines take the collection of possible adaptation options as the starting point of their analysis.

The main advantage of the first-generation assessments is that they provide a straightforward methodology (see Fig. 1.3). In addition, they provide a clear overview of the impacts of climate change for a specific sector or region. The problematic parts of the first-generation assessment methods include the step from impact assessment to the formulation of adaptation options, evaluation and policy implementation. There are two reasons for this. First, technical experts are apparently unable to effectively communicate the main results to the stakeholders. Second, the stakeholders (e.g. farmers, policymakers) are not involved in the first steps of the assessment, or have no specific interest in these steps, because these steps (scenarios development, modelling) are quite technical in nature. Therefore, the problems that the stakeholders encounter are not specifically dealt with during the studies. Thus, the 'solutions' the studies offer are not necessarily appropriate solutions to the problems of the stakeholders.

Examples of these difficulties can be found in the sections about the studies in Bolivia (Section 2.2), Senegal (2.10) and Suriname (2.11) and to a lesser extent in Colombia (2.3), Ghana (2.6), Mali (2.8) and Vietnam (2.12). The programme coordinator in Bolivia, Oscar Paz, describes the challenges as follows:

> It is evident that from the perspective of local response capacities, the results based on the use of simulation models must be complemented with studies and methodologies on the different institutional adjustments to face the impacts of the climate change ... There have been proposals to continue with the studies made in this first stage, with the inclusion of the evaluation of local capacities and the extension of the geographic scenarios to validate and to complement the obtained data.
> (Subsection 2.2.5)

In theory, the second-generation approaches could overcome the main disadvantages mentioned above. However, although the NAPA guidelines outline some 'guiding elements', they fall short of providing a structured framework (UNFCCC, 2004b). For example, they do not provide a step-by-step methodology that can be applied by a team of sector specialists or a national climate study centre. The strong point of the NAPA approach is the focus on the current capacity to deal with current extreme weather events. If vulnerable local communities are able to strengthen their capacity to deal with these events, they will probably also be better prepared for climate change. A possible weak point is that it may be more difficult to get access to 'climate funds', i.e. development programmes that are earmarked for adaptation projects as agreed in the UNFCCC or Kyoto Protocol, because the NAPA approach focuses on *current* climate events. This is discussed in more detail by Verheyen (2003).

Hence the APF may be more suitable for studies or projects like the ones described in this book. This framework is particularly applicable where the integration of adaptation measures into broader sector specific policies, economic development, poverty reduction objectives or other policy domains is desirable. The challenge, however, is to integrate the 'top-down' approach, where national plans and policies need to be formulated based on climate and socio-economic scenarios, with the 'bottom-up' approach, where local communities participate in improving their adaptive capacities. See Subsection 4.7.4 for a more detailed discussion about this subject.

4.5 Capacity Building and Awareness Raising

Ralph Lasage[14] and Michiel van Drunen[15]

4.5.1 Introduction

Important goals of the NCCSAP were to increase knowledge on climate change in the partner countries and to raise awareness of the possible impacts of climate change. To assess if these goals were achieved and to learn from the experiences during the programme, a questionnaire was sent to the involved researchers and policymakers. A different questionnaire was formulated for each group. For policymakers, the emphasis was on the incorporation of research outcomes in policy. For researchers, the emphasis was on the evaluation of capacity building. Twenty-four researchers and six coordinators/policymakers returned the questionnaire. The results are summarized in the Appendix. In the first section the results of the questionnaires completed by policymakers are listed and evaluated. The second section deals with the results of the researcher questionnaires. Subsection 4.5.2 presents the overall conclusions on the basis of the questionnaires. The respondents came from the following countries: Colombia, Ecuador, Kazakhstan, Mali, Mongolia, Senegal, Suriname and Zimbabwe.

4.5.2 Concluding remarks on the questionnaires

The results of the questionnaires lead to the following concluding remarks.

The coordinators and policymakers were very positive about the NCCSAP. They value the NCCSAP more highly than other donor-funded activities. They indicated that (participation in) the NCCSAP studies increased their awareness of climate change. In the future they would like to connect climate change related projects to disaster preparedness programmes, and – to a lesser extent – to poverty alleviation and sustainable development programmes. They also want to involve policymakers in these projects, especially those from the ministries of agriculture, waterworks, environment and economic affairs.

Most researchers worked at governmental institutions or universities. They were usually engaged in climate change impact assessments and they considered the methodologies applied suitable for the research questions they were addressing. They indicated that they increased their awareness and capacities regarding impact assessment methodologies, vulnerability assessments, adaptation strategies and climate scenario development. The research was considered to be of high quality. Nevertheless, they wanted to learn more about impact assessments and adaptation strategies, specifically in regional workshops and local

---

[14] Institute for Environmental Studies, Vrije Universiteit, De Boelelaan 1087, 1081 HV Amsterdam, The Netherlands. Tel.: +31-20-5989 506, Fax: +31-20-5989 553, Email: ralph.lasage@ivm.falw.vu.nl
[15] See Note 1, Chapter 1.

training sessions. They were quite satisfied with the technical assistance provided in the NCCSAP, but they prefer to be involved in the choice of which technical consultants will assist them in future projects. The cooperation within the projects was usually good.

4.6 NCCSAP in Comparison with Other Country Study Programmes

4.6.1 Introduction

Besides the NCCSAP, several other programmes have run with similar objectives. Below, we discuss the US Country Studies Program (US CSP; Subsection 4.6.2) and the UNEP Country Studies on Climate Change Impacts and Adaptation Assessments (Subsection 4.6.3). These two programmes were chosen because they ran at approximately the same time as the NCCSAP, because countries from different continents were involved and because the main results have already been published. We discuss the objectives, the methods and the results, and we focus on the impact, vulnerability and adaptation studies since the methodologies for emission inventories and mitigation assessments are relatively straightforward.

In Subsection 4.6.4 we compare the weaknesses and strengths of the US CSP, the UNEP Country Studies and the NCCSAP and give recommendations for future country programmes aiming at effective climate change policies in developing countries.

There are several other climate programmes for developing countries that are not discussed here. We list them briefly to assist readers interested in further information. The Climate Protection Programme for Developing Countries, run by the German *Gesellschaft für Technische Zusammenarbeit* (GTZ) focuses on integrating adaptation into development cooperation (GTZ, 2005). This programme started in 2005 and is a concrete follow up of the joint publication *Poverty and Climate Change* (African Development Bank et al., 2003). Most other climate related GTZ projects deal with mitigation.

Global Environmental Facility (GEF) support appeared to be for studies that were renamed to take advantage of additional funds around the UNFCCC. UNEP-Risø and the Japan International Cooperation Agency (JICA) focused on GHG emission inventories and mitigation assessments, with particular attention to technology transfer (Blaas et al., 2001).

The Asian Development Bank (ADB) initiated a climate change adaptation programme for the Pacific, financed by the government of Canada, which focuses on adaptation through risk reduction (Ponzi, 2002). The programme published a framework for mainstreaming adaptation (ADB, 2003), which describes how climate change impacts should be dealt with in sectors that could be affected (see also Subsection 4.7.4), but the ADB website has not yet published any other project results.

Finally, the United Nations Institute for Training and Research, UNITAR, organized workshops and training sessions in its Climate Change Training Programme (CC:TRAIN). It currently runs the Climate Change Programme (UNITAR, 2005). The Climate Change Programme focuses on National Adaptation Programmes of Action (NAPAs, see also Subsection 4.4.3).

4.6.2 US Country Studies Program

The US CSP was first announced in 1992 and it operated until 2000. The main goal of the programme was 'to enhance the capacity of developing and transition countries to conduct their own studies and develop country-specific strategies and priorities for understanding and addressing the problems of climate change' (US CSP, 1999: 1). The 56 countries involved received assistance in preparing emission inventories, evaluating mitigation options, assessing vulnerability, developing national action plans and assessing technological needs.

In the US CSP approximately 3000 people were formally trained. Analytical tools, models and libraries were installed in key ministries of each country, of the 56 US CSP study coordinators over 20 went on to become Ministers of Energy, Environment or Natural Resources of their home country, about one quarter of the UNFCCC Secretariat staff in Bonn, during the late 1990s, were US CSP graduates (Dixon, personal communication 2005).

The US CSP used four primary guidance documents: the *Revised 1996 IPCC Guidelines for National Greenhouse Gas Inventories* (Houghton et al., 1997), *Greenhouse Gas Mitigation Assessment* (Sathaye and Meyers, 1995), *Vulnerability and Adaptation Assessments* (Benioff et al., 1996) and *Steps in Preparing Climate Change Action Plans* (Benioff and Warren, 1996).

The vulnerability studies (49 countries) used the following approach (Benioff et al., 1996; US CSP, 1999; UNFCCC, 2004b):

1. Define scope of assessment process.
2. Select socio-economic, environmental and climate change scenarios.
3. Conduct biophysical and economic impact assessments and evaluate biophysical impacts of adaptive adjustments.
4. Integrate impact results.
5. Analyse adaptation policies and programmes.
6. Document and present results to decision makers and the scientific community.

Technical advisors, mainly from the USA, provided training, software and data through workshops and visits. The experts also provided guidance and review throughout the process. The studies were conducted in the following sectors: agriculture, coastal resources, fisheries, forests, grasslands/livestock, human health, water resources and wildlife. In the studies, models similar to those of the NCCSAP were used for the climate scenarios and the biophysical economic impact assessments.

The main results are summarized by Smith and Lazo (2001):

> The studies found that sea level rise could cause substantial inundation and erosion of valuable lands, but, protecting developed areas would be economically sound. The studies showed mixed results for changes in crop yields, with a tendency toward decreased yields in African and Asian countries, particularly southern Asian countries, and mixed results in European and Latin American countries. Adaptation could significantly affect yields, but it is not clear whether the adaptations are affordable or feasible. The studies tend to show a high sensitivity of runoff to climate change, which could result in increases in droughts or floods ... The major contribution of the US CSP was in building capacity in developing countries to assess potential climate impacts.

Smith and Lazo (2001) concluded that the US CSP studies in most cases focused on identifying sensitivities of systems, i.e. first-order biophysical effects. Adaptability was assessed only for coastal resources and in some of the agriculture, forest and water resources studies. Many of the studies did not analyse the implications of biophysical impacts of climate change on socio-economic conditions, cross-sectoral integration of impacts, autonomous adaptation or proactive adaptation.

They recommended that future work should attempt to develop capacity in developing countries to conduct more integrated studies of climate change impacts.

The US CSP was more or less succeeded by USAID's Climate Change Assistance Initiative, but the number and scope of projects, and the ambitions of this initiative, appear to be rather limited compared to those of the US CSP. Most activities relate to energy efficiency improvements, carbon sequestration and impact studies (USAID, 2005).

4.6.3 UNEP Country Studies on Climate Change Impacts and Adaptation Assessments

The UNEP Country Case Studies on Climate Change Impacts and Adaptation Assessments ran from February 1996 to September 1998. The main goal of the studies was to develop, test and improve methodologies to assess potential climate change impacts and evaluate adaptation strategies (O'Brien, 2000; GEF, 2005).

This GEF-financed project, implemented by UNEP was designed to:

- improve understanding of the vulnerability of countries, regions and sectors to climate change and the socio-economic impacts;
- improve regional predictions of climate change impacts;
- improve understanding of the effectiveness

and socio-economic impacts of measures to adapt to climate change;
- provide input to long-term investment planning.

The project applied and tested the *IPCC Technical Guidelines for Assessing Climate Change Impacts and Adaptations* (Carter et al., 1994) and various methods for assessing climate change impacts and adaptations under field conditions through a set of country case studies. The countries studied included Antigua and Barbuda, Cameroon, Estonia and Pakistan. As part of this project, local institutions and experts were both involved in and received training, thereby strengthening institutions and enhancing capacity to address climate change-related issues in their countries. The Centre for International Climate and Environmental Research (CICERO) provided technical assistance (O'Brien, 2000).

The UNEP Studies resulted in the *Handbook on Methods for Climate Change Impact Assessment and Adaptation Strategies* (Feenstra et al., 1998). The handbook translated the IPCC guidelines into a practical description of methods for impact and adaptation assessments and was especially designed to take into consideration the needs and requirements of developing countries (O'Brien, 2000; see also Subsections 1.5.3 and 1.5.4). Draft versions of the methods described in the handbook were applied and tested by the four country teams.

The country teams were able to pursue several methodological approaches to creating climate change scenarios, including General Circulation Model (GCM) scenarios, analogue scenarios and synthetic scenarios. Some countries, particularly Antigua and Barbuda, found the GCM scenarios of limited use, because the land area is not resolved within the grid squares of the model. Some of the teams criticized the 100-year time frame, as the uncertainties associated with the projections give the scenarios little practical meaning. The four teams used various biophysical models, economic models, empirical analogue studies and expert judgements. Researchers in Antigua and Cameroon concluded that the methodologies in the handbook were too presumptive regarding data availability (O'Brien, 2000).

The studies indicated an increase of water scarcity in Pakistan, Cameroon, Antigua and Barbuda, mostly because of increasing evapotranspiration. In Estonia the effects of climate change were considered to be manageable; the effects on agriculture and forestry were even considered beneficial. Another important conclusion was that many present-day activities increase vulnerability to climate change instead of promoting greater resiliency. Examples include development of tourism in Antigua and Barbuda, urbanization of low-lying areas in Douala, Cameroon, and increasing demand for fuel wood in Pakistan (O'Brien, 2000).

The country studies also suggested that change in climate variability and extreme events are often of greater concern than a change in average temperature or precipitation. Examples are effects of hurricanes in Antigua and Barbuda and increased risk of river floods in Pakistan. Addressing current problems is often seen as an appropriate way to increase overall resilience to climate change. Economic reforms, policy changes, improved management and increased monitoring are considered to be important means of addressing long-term climate change. Most of these measures are also beneficial in the absence of climate change! Nevertheless, these adaptation options require strategic actions as few will occur autonomously. Adaptation measures based on technological solutions were seldom identified (O'Brien, 2000).

The country studies programme was thoroughly evaluated. Glantz (1998) concluded that the project produced high quality reports, that national capacities for addressing climate change impacts and adaptation assessments have been established, and that awareness has been created among policymakers and the general public. He also concluded that the IVM/UNEP handbook (Feenstra et al., 1998) 'was very useful in the launching of these assessments and the comments of the study teams on and inputs to the draft of the Handbook have enhanced the value of the Handbook for use in future assessments'.

Among the recommendations formulated by Glantz (1998) were the following:

- The country coordinators should be asked to prepare an assessment of the strengths and

weaknesses of the methodological approaches they undertook.
- There is a need for a pre-project phase to identify the special needs of the newly assembled country teams.
- It would be beneficial to organize a workshop to bring all country study teams together and share views on the structure, process and content of their assessments.

4.6.4 Comparing the programmes

To make a proper comparison between the NCCSAP and the US and UNEP programmes it is necessary to take into account their specific features. These are summarized in Table 4.2. As can be concluded from this table, the NCCSAP and the UNEP studies are small compared to the US programme. The two younger programmes benefited from the experience gained in the US programme, regarding both methodology and programme management. The programme managers exchanged information regularly, and they jointly organized the workshop 'National Assessment Results of Climate Change: Impacts and Responses' held in San José, Costa Rica in March 1998. The recommendations by Glantz (1998) cited in Subsection 4.6.3 were taken into account in the NCCSAP, as can be read in this book.

The methodologies applied in the three programmes discussed here are not very different. The method that was applied in the US CSP (Benioff et al., 1996) built on the *IPCC Technical Guidelines* (Carter et al., 1994) and the Seven Steps Method applied in the coastal zone studies of the NCCSAP was a predecessor of these guidelines, which focuses on the effects of SLR. The IVM/UNEP handbook mentioned both Benioff et al. (1996) and Carter et al. (1994) as 'antecedents' (Feenstra et al., 1998: xxiv). The IVM/UNEP handbook offers a wider range of models, addressing a wider range of needs and a wider range of potential users, but the assessment *approach* is more or less the same as in

Table 4.2. Overview of features of the NCCSAP, the US CSP and the UNEP studies.

	NCCSAP	US CSP	UNEP studies
Budget	5,000,000[a]	US$45,000,000[b]	US$2,000,000[c]
Number of countries	13	56	4
Period	1996–2005	1992–2000	1996–1998
Aim	Capacity building	Capacity building	Methodology development
Studies	Emission inventories, Mitigation assessments, Adaptation assessments	Emission inventories, Mitigation assessments, Adaptation assessments	Adaptation assessments
Methods used for adaptation assessment	IVM/UNEP handbook IPCC Seven Steps Method	Benioff et al. (1996)	IVM/UNEP handbook
National Communications based on studies	6	11[d]	Not known
Regional workshops	2	>20	–
Global workshops	2	10	1
Visits by technical experts	>70	>100	Unknown
Programme summary	This book	US CSP (1999)	O'Brien (2000)
Programme evaluation	Blaas et al. (2001)	Evaluations of political, technical and scientific results[d]	Glantz (1998)

[a]Blaas et al. (2001); this equalled US$4,500,000 in 2000.
[b]UN (2002: 41).
[c]GEF (2005).
[d]R. Dixon (personal communication 2005).

Benioff et al. (1996) and Carter et al. (1994). Very different – second-generation – approaches were discussed in Subsections 4.4.2 and 4.4.3.

Blaas et al. (2001) evaluated the NCCSAP for the Netherlands Ministry of Foreign Affairs. They visited Bolivia, Costa Rica, Mongolia, Senegal and Vietnam (pre-NCCSAP) and they studied the reports and other products from the country teams. They concluded:

> The NCCSAP provided for an expanded study base to inform the National Communications. In all countries in which fieldwork was undertaken there had been a previous level of US country studies and GEF support. The US country studies tended to be brief overviews, completed by US experts, giving little ownership to national authorities.

Paz (personal communication 2005), who was the coordinator of both the US CSP and the NCCSAP studies in Bolivia, did not acknowledge this conclusion. He felt both programmes strengthened local capacities and he considered both programmes equally important regarding their contribution to the initial National Communication. Paz preferred the methodology of the NCCSAP because the NCCSAP experts actually visited the experts in Bolivia and worked with them. But he also mentioned that 'the technical assistance from US CSP was more complete because the international experts were mostly ready to solve the problems'.

About the cost effectiveness, Blaas et al. (2001) wrote:

> The cost effectiveness of the studies seems high, although [only] anecdotal evidence is available to the evaluation team to compare costs with US country studies and the GEF programme. The cost effectiveness has been high for several reasons including:
> - A low management charge from IVM to DGIS;
> - Significant use of national as opposed to overseas consultants in the study programme;
> - The use of overseas consultants drawn from Netherlands public service whose charges have been, in the past, lower than those of [commercial] consultants.

The relatively high number of visits by technical experts in the NCCSAP could be an indicator of the cost effectiveness. In addition, the administrative and time-keeping requirements, which were much higher in the NCCSAP than in the US CSP (Paz, personal communication 2005) may have led to a relatively high cost effectiveness, because the local researchers were forced to produce the agreed deliverables in time. On the other hand, these requirements also led to more bureaucracy: the country coordinators had to report time sheets, cost statements and interim reports every 3–6 months.

Because of the methodologies chosen, the NCCSAP, the US CSP and the UNEP programme have two common weak points that are closely connected:

- They do not perform very well at formulating effective adaptation options based on the impact studies; and
- They do not perform well on integrating the study results from different sectors and formulating integrated policy plans.

Possibly the second-generation methodologies discussed in Subsections 4.4.2 and 4.4.3 can overcome these weak points, as was discussed in Subsection 4.4.4.

4.7 Recommendations

4.7.1 Objectives

As mentioned in Subsection 1.2.1, the three main objectives in NCCSAP-I were:

- to enable developing countries to implement commitments under the Framework Convention on Climate Change;
- to create a greater awareness of climate change issues; and
- to increase the involvement of policymakers, scientists and the general public.

It can be concluded from the preceding chapters that these objectives were met to a large extent in most countries. This was also acknowledged by the evaluation that was commissioned by the Netherlands Ministry of Foreign Affairs (Blaas et al., 2001). However,

Blaas et al. also formulated two important critical remarks:

- the weakness in the adaptation studies where the focus was not on livelihoods but on physical production;
- the inability of national ministries to take involvement significantly beyond their own stakeholder group or to raise greater awareness of the climate change issue, particularly with the general public.

The latter remark was also formulated during the NCCSAP Workshop held in 2000 in Amsterdam (Dorland et al., 2001). Here, one of the discussions specifically focused on the question: 'How can the study results be successfully translated into policy plans?'. It became clear that this depends on the specific circumstances in the countries. Several requirements were, however, valid for most situations. These are elaborated in Subsection 4.7.2. The livelihood issue is addressed in Subsection 4.7.3.

4.7.2 Involvement

Most of the project members in the NCCSAP were technical experts. They were experts in e.g. climate scenarios, crop growth models or coastal zone management. Nevertheless, the results of their studies were meant to inform policymakers and to influence all kinds of policies. In order to effectively influence policymakers, they need to be involved in the research projects. Workshops provide an excellent platform for stakeholder involvement. However, it is essential to address the questions policymakers are interested in. Based on several workshops held within the NCCSAP, the participants in the Amsterdam workshop (Dorland et al., 2001) formulated the following checklist to promote policies that address the effects of climate change.

1. Policymakers need clear pictures. Therefore, the study conclusions must give transparent information. Questions that have to be addressed include:
- Which scenarios are studied?
- What are the costs and benefits in these scenarios (e.g. of certain adaptation options)?
- Which stakeholders win and lose in these scenarios? How much do they win or lose?
- What is the relative meaning of the outcomes (e.g. in terms of share of GNP)?
- Which options have priority? Why?
- Who can contribute to the realization of the plans (banks, international organizations, bilateral or multilateral programmes)?
- On which time scales do effects occur?
- Are there relevant secondary effects related to e.g. security, sustainability or health?

2. Policy actions must be taken by certain institutions. Often mitigation options must be addressed by the private sector and adaptation options by government institutions. Since governments are organized in sectors (e.g. water management, agriculture, environment, energy), it may be necessary to focus the study results towards steps to be taken in certain sectors. Such steps can then be included in the 'normal' sector policy plans. It should, however, be taken into account that measures in one sector do not interfere with effects in other sectors. For example, increased irrigation may not be feasible if the water resources are limited.

Furthermore, a single institution should be responsible for and take the lead in the implementation of the plans. It is important that such an institution is identified in the start-up phase of the studies. The study results will also be accepted sooner if the key institutional and local stakeholders (e.g. farmers, energy companies or fishermen) are directly involved in the studies. Finally, funding plays an important role. Possible funding organizations should be involved in the earliest possible phase.

3. In several countries (e.g. Colombia, Ecuador) workshops were organized to inform local stakeholders and policymakers on climate change issues. This resulted in an increased awareness of the possible impacts of climate change and resulted in a higher priority being placed on the proposed action plans in the political agenda. If the general public is well informed, political pressure on politicians can enhance the attention paid to climate change issues in the political arena.

4.7.3 Livelihoods

Besides the issue of involvement, Blaas et al. (2001) raised the issue of livelihoods:

> In the impact and adaptation assessments, participants were largely from agronomy backgrounds and did not reflect local family farm production systems. In the coastal zone workshops emphasis was placed on physical infrastructure to combat sea level rise rather than the vulnerability of existing livelihoods.

And furthermore:

> The vulnerability and adaptation studies were driven by climate change modelling ... Most importantly all models drove from impact of climate change to biophysical vulnerability, the basis of the vulnerability modelling. Little attention was paid to people – to vulnerability of livelihoods.

With hindsight, it can be concluded that in the NCCSAP livelihoods played only a minor role for two reasons. First, the studies focused on impact assessment. At the time of the studies, many debates were still going on about the very existence of human-induced climate change. Therefore, many studies were using the climate scenario study outcomes to put the issue of climate change on the policy agenda. The precise effects on livelihoods were therefore considered of secondary importance.

Second, the methodologies to assess the influence of climate change on livelihoods were lacking. The existing methodologies were very much top-down approaches that started with climate scenarios and socio-economic scenarios and although they recommended involving relevant stakeholders, the researchers concerned (climate specialists, sector specialists) were usually not equipped with the skills and tools to effectively put this into practice.

4.7.4 A suitable framework

The Amsterdam Workshop (Dorland et al., 2001) made clear that policy implementation was not included in the methods that were used in the NCCSAP studies. However, the Amsterdam Workshop did not provide a comprehensive framework either. It was also shown that the studies usually took a top-down approach in which livelihoods played only a minor role.

The Adaptation Policy Framework (UNDP, 2004) probably provides a framework that can narrow the gap between the study projects and policy implementation. Furthermore, it addresses the issue of livelihoods, although the term is usually mentioned in the context of local stakeholder involvement in the APF framework. Existing tools (such as multicriteria analysis and cost–benefit analysis) and newly developed tools (such as participatory Geographic Information Systems) would be particularly useful in this respect.

This does not necessarily mean that the first-generation approaches are already obsolete. To prepare for climate change at least the two following issues must be addressed. The first issue concerns the national and international policy agenda. Climate change affects many sectors and policy fields. Obvious examples include energy, agriculture, water resources and coastal defence. In all policy decisions made in these sectors and fields, the impacts of climate change should be taken into account.[16] The first-generation approaches provide excellent toolkits to assess these impacts, in terms of SLR, temperature changes, changes in rainfall patterns, and subsequent effects in the sectors such as flood occurrence, crop yield changes and electricity demand.

Furthermore, the first-generation approaches provide results on a relatively large geographical scale, such as a country's coastal zone, a river basin or a province. Typically, second-generation approaches consider local communities, as it is expected that adaptation also needs to be taken care of at this level.

---

[16] The term used for this is 'mainstreaming'; see Huq et al. (2003) and Bouwer et al. (2004) for more information.

Since it is impossible to gather climate change vulnerability information from all communities in a country within a reasonable time, the first-generation approaches need to be applied to obtain countrywide data.

The second issue concerns the livelihoods of people and local communities. Communities that are likely to be affected by climate change must be made aware of the risks and they must be prepared to deal with climate change impacts. It was shown in several cases that if communities are resilient enough to deal with present extreme weather events, they also have enough adaptive capacity to deal with future climate change, supposing that their socio-economic circumstances do not deteriorate too much. Therefore, strengthening the capacities of local communities to deal with extreme events is probably the key to being prepared for the impacts of climate change.

Thus, it can be concluded that first- and second-generation approaches both have distinct objectives and outputs. Therefore, they can peacefully coexist, depending on the questions that need to be addressed. Moreover, both approaches can provide added value for each other. For instance, the climate scenarios and socio-economic scenarios can be included in the vulnerability assessment of a local community. Also, local information about adaptive capacity can provide valuable information for assessing the regional impacts of, for example, floods.

It should be noted that different frameworks require different specialists. While the first-generation frameworks mainly rely on climate specialists and technical sector specialists, the second-generation frameworks call for not only local specialists and other stakeholders, but also social scientists such as economists and policy experts.

4.7.5 Conclusions

The pre-NCCSAP and NCCSAP study programmes were successful in the following fields:

- Results. The country studies revealed important and relevant results that were generally considered scientifically based.
- Approaches. The rationale of the programmes was that researchers in the countries carried out the studies with (remote) assistance of technical experts. This approach appears to have worked very well in most countries. The IPCC Seven Steps and IVM/UNEP methodologies were applicable and useful in most countries.
- Capacity building. The researchers indicated that they increased their awareness and capacities regarding impact assessment methodologies, vulnerability assessments, adaptation strategies and climate scenario development. However, it seems that in most countries neither a significant share of the policymakers nor a significant share of the general public became more aware of climate change issues as a result of the (pre-)NCCSAP study results.

The CDM is expected to become the most important financial source for climate mitigation activities in developing countries in the near future. Policy uncertainty is, however, the largest barrier for CDM projects.

New approaches for assessing vulnerability and for formulating and implementing adaptation options are required. The UNEP Adaptation Policy Framework could provide a framework for projects in these fields. However, the first-generation approaches are still needed, especially for assessing potential climate change impacts on relatively large geographical scales.

The challenge is to combine regional first-generation projects with local second-generation projects on similar subjects. A good approach would be to run such projects in parallel and to organize common workshops that stimulate the researchers of both projects to cross-fertilize each other's work.

Appendix
Questionnaire Results

Policymakers and coordinators

In total, six of the country coordinators or other policymakers who were involved responded to the questionnaire sent by the NCCSAP management. Their reactions are summarized below.

In what type of organization are you employed?

I work for a	Respondents
Government agency	2
Other governmental organization	3
University	1

All of the policymakers who returned the questionnaire worked at government agencies or other governmental organizations. None of the reactions came from ministries.

What was your affiliation to the (pre-)NCCSAP Phase I?

I was/am a	Respondents
National focal point coordinator to the UNFCCC	1
Country coordinator	4
Team leader	1
Other (please identify, e.g. project coordinator)	1

One of the respondents was a national focal point and country coordinator. Most respondents were country coordinators.

Did your involvement in the NCCSAP lead to an increase of your awareness of climate change issues? If yes, please describe in what way.

All the respondents replied that their awareness of climate change issues had increased. This increase was caused by the increase in knowledge in the overall field of climate change. Some learned how climate change scenarios work, others increased their knowledge of mitigation and adaptation options. Many respondents indicated that they improved their knowledge of the impact of climate change on their country as a whole and on the vulnerability of certain areas, such as the coastal zone, and certain sectors, such as the economy, agriculture and ecosystems. The development of adaptation and mitigation strategies and the way to implement them gave insight into how they can deal with the effects of climate change. One respondent even worked on the preparation of a national action programme on climate change covering all the above.

Would you like to be involved in future climate change adaptation research activities?

All the respondents answered positively to this question; they expected to gain more knowledge in future climate change related research. Climate change research is a very dynamic area and many new insights are being developed; the respondents wanted to keep up with these developments. Some indicated they needed to continue projects that had started in the first phase of the NCCSAP. Most respondents had become more aware of vulnerable sectors or regions in their country and they wanted to study possible adaptive measures to ameliorate the impacts of climate change. A remark made by two respondents was that knowledge on climate change needs to be shared with national and local authorities. This was not realized in the first phase.

Could climate change adaptation research activities be made more useful for your country by linking the research directly with poverty alleviation, sustainable development, natural disasters or conflict and violence?

All the respondents considered the link to natural disasters important, because of the immediate effects on the community of a drought or flood.

Four respondents considered poverty alleviation to be a good link because developing countries are more vulnerable to climate change impacts. As a country develops, it can improve its climate change policy and the population will become less vulnerable. Another respondent stated that a link with poverty alleviation would help to draw public sector institutions' attention to climate change.

Four respondents considered sustainable development to be a good link. In one instance, this was because the country (Mali) lies in an arid/semi-arid region. The stocks of natural resources, such as water, are very limited in comparison to the demands of the population. These resources need to be managed in a sustainable way. Another respondent stated that sustainable development is the key focus of the government in his country; hence a link will increase the support for climate change research and policy.

Only one respondent chose the link to conflict or violence. In his country limited resources, such as pastures, water and agricultural land, are a source of conflict. Perhaps the development of an integral adaptation strategy could help contain these conflicts.

Do you think it is useful if policymakers participate actively in the projects from the start?

Five respondents reacted positively and one negatively. For the respondents, the main reasons to actively involve policymakers in the projects is that the project results will then be used more quickly for policy making. One of them wrote: 'They [policymakers] also need insight in the technical aspects. This will remove some barriers to successful implementation of the suggested options.' Another advantage is that the goals of the project will match the national development priorities. The involvement of policymakers could also help to make the output of the project more action and policy oriented. The respondent who replied negatively to this question argued that policymakers should use the outcomes of the project. It seems he aims at the same goal as the other respondents, but in his opinion it is possible to realize it in another way.

Is climate change an important issue for your country?

All respondents replied that climate change is an important issue for their country, because of socio-economic vulnerability, the effects on resources (water and food), the effects on regions (coastal zone), the increase in variability and the locations of the countries (low-lying, arid, semi-arid, etc.).

What did you think of the approaches used in the NCCSAP?

The approach used in the NCCSAP	Yes	No	Remarks
was simple and straightforward	4	2	Easy to follow
was too complicated	1	3	
was applicable	3	1	
was based on IPCC guidelines for GHG emissions	4		
was based on the IVM/UNEP handbook on adaptation	5		
was the IPCC 7 steps method	1	2	Useful guide, not familiar with this
revealed new information	3	1	About sea level rise (SLR)
was scientific in nature	5	1	A lot of research was done
involved relevant policymakers	4	2	No, more scientific
involved relevant NGOs	3	2	
involved regional experts	3	2	
involved other relevant stakeholders	3	2	
convinced policymakers	4	1	It resulted in a National Communication

Several respondents said they found the approach easy to follow (four) and scientific in nature (five). They said that the outcomes were convincing for policymakers (four) and that they were involved in the studies (four). The involvement of other groups is less clear, half of the respondents said that NGOs, experts and other relevant stakeholders were involved (three), but others said that they were not (two).

Do you have interest to be involved in future activities?

All of the respondents wanted to be involved in future activities. One of them kept it clear and simple, he answered 'of course'. Out of the answers to the previous questions, it became clear that respondents thought the projects were useful and that they learned from them. They would like to increase their knowledge of climate change and especially of adaptation options. This is also the outcome of the answers to questions 10 and 11, in which they rated the importance of adaptation as high.

How do your rate the interest of your organization in climate change adaptation?

All respondents (six) answered 'high'.

How do you rate the interest at the national policy level in climate change adaptation?

Five respondents answered 'high' and one 'medium'.

Which ministries and other government organizations should be involved in climate change adaptation policy design?

Ministry	Respondents
Agriculture	5
Waterworks	5
Environment	5
Economic Affairs	4
Cabinet	2
External Affairs	2
Finance	2
Education	2
Social Affairs	2
Housing	2
Tourism	2
Internal Affairs	1

Almost all of the respondents found it important that the ministries of environment, agriculture and waterworks are involved in adaptation policy design. One respondent chose all of the ministries, which has a great impact because of the small amount of responses. If we consider this, the other ministries are not as important in terms of involvement in adaptation policy design according to the respondents.

How do you rate the value of the NCCSAP and other programmes?

How do you rate the value of	Average rate[a]
the UNFCCC for your country?	3.8
the IPCC assessment reports for your country?	3.6 (2–5)
the NCCSAP for your country?	4.2
other donor funded research activities?	3.0

[a] on a scale of 1–5 where 1 = insignificant and 5 = extremely valuable

The NCCSAP and UNFCCC were highly valued. The IPCC was less valued, but most still valued it above three. One respondent gave the IPCC a low value because it is not widely known in his country. Other donor funded research activities received low values, but none of the respondents explained why.

Have the NCCSAP and other donor funded projects been useful for adaptation policy design and implementation in your country?

One respondent gave a negative reaction to this question. In his country, adaptation was not studied in the project. According to the policymaker, it is urgent that projects and programmes for the development of adaptation strategies are included in the second stage of the programme. In other countries, national adaptation programmes had been developed and adaptation strategies were formulated according to the UNFCCC framework. In one country, adaptation measures for the water infrastructure were developed. This programme increased public awareness of climate change and helped with designing and implementing adaptation policy.

Do you have the impression that policymakers consider the results of NCCSAP biased because the studies were financed by The Netherlands? Please explain

Two respondents had no opinion on this subject. The other four did not see the results as biased. One of them wrote, 'The results were considered as the work of the research team funded by the NCCSAP, not as a result of the NCCSAP.' There is no indication that the results were considered biased.

Do you think the results of the project were useful for adaptation policy development and implementation?

Three respondents had no opinion on this subject. The other three responded positively. They said this was because they had prepared National Communications and action programmes, because the outcomes of the projects could be used by city planners and ecosystem managers, and because the outcomes could be used for the development of water infrastructure policies.

Do you have other comments or suggestions for improving future donor activities?

The respondents had the following suggestions to improve future donor activities:

- Address pilot studies at community level to evaluate impacts on the development of the economy.
- It is desirable that the cooperation is continued in the future.
- It would be useful if the NCCSAP could be involved in the funding of new adaptation projects on the coastal zone.
- There is a need for a global framework to exchange knowledge on different policies, even though the initiatives should be locally driven. Through this framework, external support can be found and researchers can communicate with each other.

Researchers

Over 100 local researchers were involved in the NCCSAP. In total, 24 researchers returned the questionnaires and their answers are summarized below.

In what type of organization are you employed?

Employer	Respondents
Ministry	1
Government agency	5
Other governmental organization	7
University	7
Consultancy firm	–
Other NGO	1
Other (please identify)	3

Most respondents were government employees, 13 in total. Only one of these works for a ministry, the others work at agencies or other government organizations. Universities were also a big supplier of respondents; seven of the reactions came from people who work in academic institutions. None of the consultants involved in the NCCSAP returned a questionnaire.

Did your involvement in the NCCSAP lead to an increase of your awareness on climate change issues? If yes, please describe in what way

Subjects mentioned in response to this question	Number of mentions
Technical	
Learned the impact of CC/CV[a] (SLR, extremes)	7
Impact of CC on country	5
Learned effect of CC on coastal zone management	3
Global impact of CC	3
Development of adaptation strategies	3
Knowledge on climate scenario development	2
Legal and institutional aspects of CC	1
Development of GHG emission inventory	1
Communication	
First time research on CC/CV	3
Learned new research methods and approaches	2
Raised awareness on CC	2
Now know that many parts of CC research are based on assumptions	2
Identified socio-economic factors that drive CC	1
No increase	
Participated in IPCC since 1990	3

[a]CC/CV, climate change/climate variability.

Almost all of the respondents said that their awareness of climate change issues increased during the programme. Twenty-one reacted positively to this question. The table lists the subjects they mentioned, including the impact of climate change (seven), the impact of climate change on a global scale (three), and the impact on a country scale (five). An interesting remark from one respondent is that he now knows that many parts of climate change research are based on assumptions. One of the three researchers who responded negatively to this question was involved in the IPCC before joining the NCCSAP; the other two were people with working experience on climate change issues.

How has capacity to assess climate change impacts increased?

Subjects mentioned in response to this question	Number of mentions
Technical/data	
Improved knowledge of methodologies	6
Increase in understanding of the aspects of CC	3
More knowledge of the technical background (models, scenarios)	3
Availability of information	2
Increase in knowledge of influence of CC on water resources	1
Communication/learning	
Gained experience in assessing CC impacts, vulnerability and adaptation	8
Increased means to conduct vulnerability assessments	5
Working with specialist consultants from other countries	2
Formulated national action programme	1
Organization/structure	
The multidisciplinary approach of the studies	3
Involvement of local authorities in the NCCSAP studies	2
Interaction with other institutions in country	1
Other	
Included adaptation in policy	1
Visualization of CC problem from different points of view	1

The respondents increased their capacity to assess climate change impact through many means. The table lists their remarks and the number of times they were mentioned. The experience of working on a climate change project (eight) was very important for increasing capacity; this includes working with specialist consultants (two), interacting with other institutions (one), using a multidisciplinary approach (three) and increasing the means to conduct vulnerability assessments (five). For one person, this led to the ability to approach the climate change problem from different viewpoints. The increase in knowledge on methodologies (six), technical background (six) and aspects of climate change (three) were also important. Some of the respondents increased their capacity through better availability of data (two).

Did the technical assistance provided in the NCCSAP satisfy your expectations?

In the opinion of the respondents, the technical assistance contributed to the increase in the capacity to perform climate related research. Six found the technical assistance crucial, 11 found it very useful and six found it useful. No respondent found it almost or completely useless. For 13 respondents the technical assistance satisfied their expectations, while ten said it partly met their expectations. Some of the respondents said they learned a great deal from the technical assistance (five), others were more precise and explained what they learned.

What were the experiences with technical assistance?

Experience	Respondents
Helped with development and practical construction of CC scenarios	3
Mathematical modelling of hydrological processes	1
Methodology guideline was very clear and applicable	2
Good when TA was in the country	3
Quality was excellent	2
Learned a lot from TA	5
Learned to work in a multinational team	1
Integrate research outcomes in policy development	1
Access to technological information and data	2
Some problems with financial assistance	1
Need to be more open to knowledge in the country	2
Need more communication in the beginning of the project	1
Expected more from TA, only one expert came	1
Communication over email did not work very well	2

Half of the respondents learned technical skills, such as construction of climate change scenarios (three), modelling (one) or acquiring technical information and data (two). The second part of the table lists negative experiences. Some of these are of an organizational nature. Others include communication (three) and how knowledge of the country was taken into account (two). It is important to consider these remarks in order to improve the cooperation and project structure for the follow-up research.

How to improve the technical assistance in the future?

Topic	Number of mentions
Communication	
Contact with other countries, visiting other groups with same subject	3
Visit sites, have workshops and summer schools	1
More constant and closer contact with TA	1
Capacity building	
Organize workshops to share experience	3

Topic	Number of mentions
Technical expert in country (training)	3
Place researchers in the Netherlands for a few weeks	2
Communicate and transfer knowledge to the country so they can perform the research by themselves when the project ends	2
Give more attention to training of national experts	1
Organization/structure	
Involve local coordinators in designing projects	2
Give attention to continuity in cooperation	2
Structured project guidelines	1
More flexible approach in project design	1
Base proposal on expert knowledge	1
Work plan in beginning of project	1
Involve stakeholders from the start	1
More budget to do in-depth research on aspects	1
Use local expert (who knows the area very well)	1
Simplified financial procedure would improve TA	1
Make room for country-driven project objectives	1
Pilot studies, to test adaptation options	1

The table contains a list of topics on which the respondents would like to receive technical assistance in future projects. It also highlights how the assistance could change in the future. The remarks are grouped under three themes, Communication, Capacity building and Organization/structure. Considering the theme of communication three respondents indicated they want to work with other countries that experience effects of climate change that are similar to the effects on their own countries. They want to share their knowledge and experiences and they hope to learn from others. The respondents made many different remarks and many remarks were made only once. Ten respondents remarked on capacity building, especially the placement of a technical expert in the country (three) and the placement of researchers in the Netherlands (three). Their ultimate goal was to be able to perform climate research themselves after the programme had ended (two). Remarks listed under organization included some respondents who wanted greater involvement of local people and to take the goals of the country into account in the research.

What kind of training or technical assistance would you like to receive in the future?

I would like to receive assistance/training in	Respondents
Vulnerability and adaptation methodologies	17
Climate change scenario development	14
Integrated risk management	12
Multidisciplinary research	11
Geographic information systems	8
Socio-economic scenario development	8
Physical modelling	7
Policy plan design	6
Statistics	5
Project management	5
Policy analysis	4
Participatory processes	3
International law	2
Decision support/cost–benefit analysis	2
Financial administration	1

More than two-thirds of the respondents would have liked to receive assistance in vulnerability and adaptation methodologies (17). They also found climate change scenario development important (14). Training in integrated risk management, which is linked to vulnerability and adaptation, was also considered to be an important subject (12). Interest in research methods was moderate, eight wanted to learn more about geographic information systems and 11 wanted to learn to conduct multidisciplinary research. The lack of interest in policy analysis is striking, only four respondents mentioned it. One would expect this to be important to influence policy making.

How would you like to see technical assistance organized in future activities?

How do you prefer to organize technical assistance (TA)?	Respondents
I want to select the experts providing TA myself	9
I want to plan and organize the TA myself	8
I want the donor to select the experts for me	7
I want the donor to plan and organize the TA	7
I want TA by experts from my own region	7
I want TA from the best experts in the world	7

The table lists the responses to the question on how to organize technical assistance in the future. Half of the respondents wanted to plan and organize the assistance and select the experts themselves (eight) and the other half wanted the donor to do this (seven). The choice of approach depends on the country. This choice needs to be discussed in the beginning of the project.

Are you satisfied with your results of the project?

Are you satisfied with your results of the project?	Respondents
Extremely satisfied	4
After the project there was an increase in the application of ICZM	3
Wrote a good report	2
Raised policy awareness on CC	2
Satisfied, but there are new developments on CC, so need new research	2
Of course, gave opportunity to show effect of CC on coast	2
Evidence of information gaps, potential effects of CC and SLR are known to the country	1
Results from the subregion were the best, also used by policymakers	1
Identified gaps in local knowledge, which need to be filled	1
Work done was of good quality, is basis for future research	1
Raised stakeholder awareness on CC	1
First National Communication	1
Provided negotiation tools for the international CC arena	1
No follow up, results were not communicated to local stakeholders	1

All the respondents were satisfied with the results of the projects; four researchers said they were extremely satisfied. Some projects influenced policy making, such as the application of Integrated Coastal Zone Management (ICZM) (three), the production of National Communications and international negotiation tools for climate change. But most projects were less influential and prepared the country for follow-up activities. These projects identified gaps in knowledge, raised awareness on climate change (stakeholder and policymakers) and made initial attempts to study the impacts of climate change on the countries involved.

What do you think of the approach applied for the studies in your country?

The approach I used	Yes	No	N/A
was simple and straightforward	14	5	5
was too complicated	4	12	8
was applicable	20	1	3
was based on IPCC guidelines for GHG emissions	11	4	9
was the IVM/UNEP handbook on adaptation	8	3	13
was the IPCC Seven Steps Method	12	1	11
revealed new information	16	2	6
was scientific in nature	18	2	4
involved relevant policymakers	11	4	9
involved relevant NGOs	5	9	10
involved regional experts	14	3	7
involved other relevant stakeholders	12	2	10
convinced policymakers	10	2	12

All but one respondent found the approach applicable. According to 14 respondents, it was simple and straightforward. Five, however, said it was not simple or straightforward. Eighteen respondents thought the research was scientific and revealed new information (16). Many thought the NGOs were not adequately involved. Half of the researchers said the projects convinced policymakers. Some respondents indicated that they lack information and therefore cannot use the IPCC methods. There is a need to conduct more research before these methods are applicable in developing countries.

Was the cooperation in the project team satisfactory?

Remarks made about team satisfaction	Respondents
Good team and coordination, very engaged	8
Team comprised multidisciplinary investigators and experts	4
Exchange of knowledge and data within the project went very well	4
Capacity to direct research teams	1
Local coordinator quickly picked up the way to work	1
Cooperation was limited, but we wanted to	1
Qualified team	1
No experts, but supported by experts	1
Some problems with communication	1
The involved researchers did not have enough knowledge on CC, which led to long discussions	1
Appointments not kept and deadlines frequently exceeded	1

According to a number of respondents, the teams were very engaged and worked well together (eight). Responses were also positive about the exchange of knowledge within the project (four) and the multidisciplinary character of the teams (four). The negative remarks seem to be project specific, but should be considered in the follow-up research programme.

Are the results of the project useful for adaptation policy development and implementation?

Usefulness for adaptation policies and implementation	Respondents
Identification of vulnerable areas for policymakers	7
Sufficient input for coastal zone management plan	3
Developed a national action programme	3
Adaptation measures implemented	2
Identification of research needs	2
The research did a lot, however, there is room for further development and implementation of adaptation policy. Especially dissemination to policymakers and stakeholders of different economic sectors	2
Informed local policymakers on the prevention of mudflows	2
Provided useful technical basis for policy	2
Continuous awareness raising	2
First assessment of costs of adaptation	1
Capacity was built in national parties involved	1
Future planning for coastal zone	1
Fit for national scale, not for regional and local scale	1
Need to investigate physical changes more	1
Creation of CC office in the country	1
The research was provocative, it made us ask questions we had never asked before	1
Contribution of financing	1

In response to this question seven respondents noted the contribution to the development of adaptation policy consists largely of identification of vulnerable areas. Other points were the development of adaptation measures, such as national action programmes (three), prevention of mudflows (two) and technical input (two) for policies such as the coastal zone management plan (three) and the costs of adaptation (one). A few projects led to implementation of adaptation measures (two). In one country, a climate change office was created as a result of the project. Another respondent indicated that the project has led to an increase in capacity among those involved.

Two respondents said there is still much to do in the area of implementation of adaptation policy, especially on dissemination of information to policymakers and stakeholders. One respondent said the project was fit for the country as a whole, but not for regional or local policymakers. In subsequent projects, more attention should be given to the local stakeholders and policymakers. One respondent wrote that the project was provocative; it made people ask questions they had never asked before. This made them aware of climate change and its impacts.

Would you be keen on working in future donor activities such as the NCCSAP?

I want to work on future activities because	Respondents
I acquired a lot of knowledge (with the first phase)	5
I can provide more information and scientific aspects for policymakers	4
I want to work with neighbouring countries with the same problems	3
I want to raise awareness	3
We could not do this kind of research by ourselves	2
There is more experience in multidisciplinary groups	2
I want to benefit from Dutch knowledge (e.g. adaptation technology)	2
Suriname is threatened by CC, more information is needed	1
There are many goals, and I want to help country in confronting SLR to coastal zone	1
Procedures are clear now, so future work will be quicker	1
I can contribute to the understanding of the natural phenomena, but also the benefit of adaptation	1
Global and local environment is dynamic and research needs to be continuous	1
Development of international contacts broadens one's view	1

All of the respondents wanted to work on future donor activities, because they learned a great deal in the first phase and wanted to acquire more information (five). The respondents wanted to profit from the Netherlands' experience with adaptation strategies (two); some indicated they are not able to do this kind of research by themselves (two). Others said that the information acquired in the projects provides policymakers with technical background for their work (four). In future projects, the respondents would like to work with neighbouring countries (three) to learn from each other. These international contacts contribute to a broadening of the view of the researchers (one).

Do you have suggestions for improving the donor activities?

Suggestion	Respondents
Allocate funds for application of the research results into practice or implementation, finance adaptation projects, increase the awareness to policymakers	3
Enhance technology transfer	1
Formulate a strategy to maintain contact when the team has finished their project; the NCCSAP cooperation should be regular	2
Make sure there is follow-up research that uses the collected data	1
Involve (local) experts at an early stage in the project	2
Technical assistance needs to be more structured with guiding documents	1
Take the priorities of the country into account and their scientific approach, in order to join efforts	1
Translate the NCCSAP documents into French	1
Give more attention to communication with stakeholders	1
Reduce the number of activity reports that have to be written	1
Better quality control on the results used for national reports	1

References

ADB (2003) *Guidelines on Adaptation Mainstreaming for Pacific Department Operations (draft).* Asian Development Bank, Climate Change Adaptation Program for the Pacific (CLIMAP).

African Development Bank *et al.* (2003) *Poverty and Climate Change – Reducing the Vulnerability of the Poor through Adaptation.* The World Bank, Washington, DC, 43 pp. Available at: http://www.worldbank.org

Becker, C.R., Breinburg, H., MacDonald, H., Playfair, M. and Ramdihansing, R. (1999) *Greenhouse Gas Emission Inventory for Suriname 1994, Project Country Study Climate Change Suriname.* Technical Report 7, Meteorological Service, Paramaribo, Suriname, 108 pp.

Benioff, R. and Warren, J. (eds) (1996) *Steps in Preparing Climate Change Action Plans: a Handbook.* US Country Studies Program, Washington, DC.

Benioff, R., Guill, S. and Lee. J. (eds) (1996) *Vulnerability and Adaptation Assessments: an International Guidebook.* Kluwer Academic Publishers, Dordrecht, The Netherlands.

Blaas, H., Keefe, P.O., Kieskamp, W.M. and Marticorena, B. (2001) *The Netherlands Climate Change Studies Assistance Programme: Evaluation of the Period 1993–1999.* ETC International and Hans Blaas Consultancy, The Netherlands.

Bolaños, R.A. and Watson, V. (1993) Mapa Ecológico de Costa Rica según el Sistema de Clasificación de Zonas de Vida del Mundo de L.R. Holdridge. Escala 1:200,000. CCT – ICE. San José, Costa Rica.

Bouwer, L.M., Dorland, C., Aerts, J.C.J.H. and Gupta, J. (2004) Adaptation and funding in climate change policies. In: Kok, M.T.J. and De Coninck, H.C. (eds) *Beyond Climate – Options for Broadening Climate Policy.* RIVM report 500019001/2004. National Institute for Public Health and Environment (RIVM), Bilthoven, The Netherlands. Available at: http://arch.rivm.nl/ieweb/ieweb/Reports/rep500019001.pdf

Broadus, J.J. Millman, S.E., Aubry, D. and Gable, F. (1986) Rising and damming of rivers: possible effects in Egypt and Bangladesh. In: Titus, J.G. (ed.) *Effect of Changes in Stratospheric Ozone and Global Climate 4.* Environmental Protection Agency (EPA) and United Nations Environment Programme (UNEP), Washington, DC.

Bygrave, S. and Bosi, M. (2004) *Linking Project-based Mechanisms with Domestic Greenhouse Gas Emissions Trading Schemes.* Organisation for Economic Cooperation and Development (OECD)/International Energy Agency (IEA) Project for the Annex I Expert Group on the UNFCCC, Paris.

Carter, T.R., Parry, M.L., Harasava, H. and Nishioka, S. (1994) *IPCC Technical Guidelines for Assessing Climate Change Impacts and Adaptations.* Intergovernmental Panel on Climate Change, Department of Geography, University College London, UK and Centre for Global Environmental Research, Tsukuba, Japan, 59 pp.

Dankwa, J.B. (1974) *Maximum Rainfall Intensity. Duration Frequencies in Ghana.* Meteorological Services Departmental note number 23, Ghana.

Delft Hydraulics (1992) *A Global Vulnerability Assessment,* 2nd revised edition. Rijkswaterstaat, The Netherlands.

Dennis, K.C., Niang-Diop, I. and Nicholls, R.J. (1995) Sea-level rise and Senegal: potential impacts and consequences. *Journal of Coastal Research, Fort Lauderdale* (Special Issue) 14, 243–261.

Diop, E.S. and Ba, M. (1993) *Les mangroves du Sénégal et de la Gambie*. Rapp. Technique du Projet Conservation et Utilisation Rationnelle des Forêts de Mangrove de l'Amérique Latine et de l'Afrique. ITTO/ISME, 22–40.

Doorenbos, J. and Pruitt, W.O. (1986) The crops water needs. *World Organisation for Food Bulletin* 24, Rome, p. 198.

Dorland, C., Willemsen, F.H., Brander, L. and van Drunen, M.A. (2001) *Netherlands Climate Change Studies Assistance Programme – Report of the NCCSAP Workshop, 27–28 November 2000, Amsterdam, The Netherlands*. IVM Report (W-01/14). Vrije Universiteit, Amsterdam, 200 pp.

Downing, T. (1992) *Climate Change and Vulnerability Places: Global Food Security and Country Studies in Zimbabwe, Kenya, Senegal and Chile*. Environmental Change Unit, Oxford, UK, 54 pp.

ECN (1999) *Potential and Cost of Clean Development Mechanism Options in the Energy Sector: Inventory of Options in Non-Annex I Countries to Reduce GHG Emissions*. December 1999, ECN-C-99-095. Energy Research Centre of The Netherlands (ECN), Petten, The Netherlands, 32 pp.

EcoSecurities CDM project. Information available at: http://www.ecosecurities.com/

Emery, K.O., Aubrey, D.G. and Goldsmith, V. (1988) Coastal neo-tectonics of the Mediterranean from tide-gauge records. *Marine Geology* 81, 281–295.

Environmental Protection Agency (2001) *Initial National Communication of Ghana*. Environmental Protection Agency, Accra, Ghana, 172 pp. Available from: http://unfccc.int

Environmental Protection Council (EPC) (2001) *Initial National Communication under the United Nations Framework Convention on Climate Change*. EPC, Republic of Yemen, Sana'a, Yemen, 72 pp.

ETC International (2004) *The Netherlands Climate Change Studies Assistance Programme*. ETC International, Leusden, The Netherlands. Available at: http://www.nccsap.net

Faye, S., Niang-Diop, I., Cisse Faye, S., Evans, D.G., Pfister, M., Maloszewski, P. and Seiler, K.P. (2001) Seawater intrusion in the Dakar (Senegal) confined aquifer: calibration and testing of a 3D finite element model. In: Seiler, K.P. and Wohnlich, S. (eds) *New Approaches Characterizing Groundwater Flow*. A.A. Balkema, Lisse, The Netherlands, pp. 1183–1186.

Feenstra, J.F., Burton, I., Smith, J.B. and Tol, R.S.J. (1998) *Handbook on Methods for Climate Change Impact Assessment and Adaptation Strategies*. Institute for Environmental Studies (IVM)/United Nations Environment Programme (UNEP), Vrije Universiteit, Amsterdam, 464 pp. Available at: http://www.vu.nl/ivm

Gaye, A.T., Dabo, E.M.F., Sall, S.M., Sambou, E. and Fongang, S. (1998) *Scénarios de changements climatiques pour des études d'impacts sur l'agriculture et les zones côtières du Sénégal*. Ministère de l'Environnement, Dakar, 16 pp.

GEF (2005) *Factsheet: Country Case Studies on Climate Change Impacts and Adaptations Assessments*. Global Environment Facility (GEF), Nairobi, Kenya, 2 pp. Available at: http://www.gefweb.org/Outreach/outreach-Publications/Project_factsheet/Global-impa-6-cc-unep-eng.pdf

Glantz, M.H. (1998) *Final Evaluation of the UNEP/GEF Project 'Country case studies on climate change impacts and adaptation assessments' (GF/2200-96-09)*. United Nations Environment Programme, Nairobi, Kenya, 34 pp. Available at: http://www.unep.org

Global Commons Institute (2004) *Contraction and Convergence Framework*. Available at: www.gci.org.uk

GTZ (2005) *Adaptation to Climate Change, Causes, Impacts, Responses*. Deutsche Gesellschaft für Technische Zusammenarbeit (GTZ), Eschborn, Germany, 8 pp. Brochure available from: http://www.gtz.de/climate

Hansen, D. and Rattray, M. (1966) New dimension in estuarine classification. *Limnology and Oceanography* 11(3), 319–326.

Holdridge, L.R., Grenke, W.C., Hatheway, W.H., Liang, T. and Tosi, J.A., Jr (1971) *Forest Environments in Tropical Life Zones: a Pilot Study*. Pergamon Press, Oxford, UK, 747 pp.

Hoozemans, F.M.J., Marchand, M. and Pennekamp, H.A. (1993) *Sea Level Rise. A Global Vulnerability Assessment*. Delft Hydraulics, Rijkswaterstaat, Delft, The Hague, The Netherlands, 184 pp.

Houghton, J.T., Meira Filho, L.G., Lim, B., Treanton, K., Mamaty, I., Bonduki, Y., Griggs, D.J. and Callender, B.A. (eds) (1997) *Revised 1996 IPCC Guidelines for National Greenhouse Gas Inventories*. Intergovernmental Panel on Climate Change (IPCC)/OECD/International Energy Agency (IEA), IPCC National Greenhouse Gas Inventories Programme (NGGIP), Japan. Available from: http://www.ipcc-nggip.iges.or.jp/public/gl/invs1.htm

Hubert, H. (1917) Progression du dessèchement dans les régions sénégalaises. *Annales de Geographie*, Paris, XXVI(143), 376–389.

Huq, S., Rahman, A., Konate, M., Sokona, Y. and Reid, H. (2003) *Mainstreaming Adaptation into Climate Change in Least Developing Countries (LDCs)*. International Institute for Environment and Development, London, 42 pp. Available at: http://www.iied.org/docs/climate/main_ldc_rprt.pdf

Hydrometerological Services (1996) *Vietnam Coastal Zone Vulnerability Assessment – Report 7*. Department of Foreign Affairs, Government of Vietnam, Final Version April 1996.

ICWE (1992) *Development Issues for the 21st Century*. The Dublin Statement and Report of the International Conference on Water and the Environment (ICWE), Dublin, Ireland, 26–31 January 1992. World Meteorological Organization, Geneva.

Institute of Meteorology and Hydrology (2001) *Initial National Communication of Mongolia*. Institute of Meteorology and Hydrology, Ulaanbaatar, Mongolia, 110 pp. Available from: http://unfccc.int

IPCC (1991) *Assessment of the Vulnerability of Coastal Areas to Sea Level Rise – a Common Methodology*. Report of the Coastal Zone Management Subgroup of Intergovernmental Panel on Climate Change (IPCC) Working Group II. National Institute for Coastal and Marine Management/RIKZ, The Hague, The Netherlands, 56 pp.

IPCC (1992a) *Assessment of the Vulnerability of Coastal Areas to Sea Level Rise – a Common Methodology*. Report of the Coastal Zone Management Subgroup of IPCC Working Group III, National Institute for Coastal and Marine Management/RIKZ, The Hague, The Netherlands.

IPCC (1992b) *Global Climate Change and the Rising Challenge of the Sea*. Report of the Coastal Zone Management Subgroup, IPCC, Geneva 2, Switzerland.

IPCC (1994) *Preparing to Meet the Coastal Challenges of the 21st Century*. Conference report World Coast Conference, Noordwijk, The Netherlands.

IPCC (1996) *Climate Change 1995. The Science of Climate Change. Contribution of Working Group I to the Second Assessment Report of the Intergovernmental Panel of Climate Change*. Cambridge University Press. (edited by Houghton, J.T. et al.).

IPCC (2001) *Third Assessment Report – Climate Change 2001*. Intergovernmental Panel on Climate Change (IPCC). Available at: www.ipcc.ch

IPCC/CZMS (1991a) *Strategies for Adaptation to Sea-level Rise*. Intergovernmental Panel on Climate Change (IPCC)/Coastal Zone Management Subgroup (CZMS), The Hague, The Netherlands, 24 pp.

IPCC/CZMS (1991b) *The Seven Steps to the Vulnerability Assessment of Coastal Areas to Sea Level Rise*. Intergovernmental Panel on Climate Change (IPCC)/Coastal Zone Management Subgroup (CZMS), The Hague, The Netherlands, 122 pp.

Janssen, M.A., Goosen, H. and Omtzigt, N. (2005) A simple mediation and negotiation support tool for water management in the Netherlands. *Landscape and Urban Planning* (in press). Available at: http://www.sciencedirect.com/

Kazakh Research Institute for Environment Monitoring and Climate (KazNIIMOSK) (1998) *Initial National Communication of the Republic of Kazakhstan under the United Nations Framework Convention on Climate Change*. KazNIIMOSK, Almaty, Kazakhstan. Available from http://unfccc.int, 75 pp.

Klein, R.J.T., Nicholls, R.J. and Mimura, N. (1999) Coastal adaptation to climate change: Can the IPCC technical guidelines be applied? *Mitigation and Adaptation Strategies for Global Change* 4, 239–252.

Kulshreshtha, S.N. (ed.) (1993) *World Water Resources and Regional Vulnerability: Impact of Future Changes*. International Institute for Applied Systems Analysis, Novagraphic Publishing, Vienna, Austria.

Lapenis, A.G., Oskina, N.S., Barash, M.S., Blyum, N.S. and Vasileva, Y.V. (1990) The Late Quaternary changes in ocean biota productivity. *Okeanologiya/Océanologie* 30, 93–101.

Least Developed Countries Expert Group (2002) *Annotated Guidelines for the Preparation of National Adaptation Programmes of Action*. UNFCCC, United Nations Development Programme (UNDP), 43 pp. Also available from: http://www.undp.org/cc/napa.htm

McCarthy, J.J., Canziani, O.F., Leary, N.A., Dokken, D.J. and White, K.S. (2001) *Climate Change 2001: Impacts, Adaptation and Vulnerability*. Intergovernmental Panel on Climate Change, Cambridge University Press, UK, 1032 pp.

Metz, B., Davidson, O.R., Martens, J.W., Van Rooijen, S.N.M. and McGrory, L.V.W. (eds) (2000) *Methodological and Technological Issues in Technology Transfer: a Special Report of the Intergovernmental Panel on Climate Change*. Cambridge University Press, UK, 432 pp. Available from: http://www.grida.no/climate/ipcc

Michel, P., Naegele, A. and Toupet, C. (1969) Contribution à l'étude biologique du Sénégal septentrional. I. Le milieu naturel. *Bulletin de l'Institut Fondamentale d'Afrique Noire (IFAN)*, Dakar XXXI(3), 756–839.

Ministry of Science, Technology and Environment of Vietnam (1992) *Vietnamese Red Book of Rare and Endangered Species*. Science and Technics Publishing House, Hanoi.

Ministry of Sustainable Development and Planning (2000) *First National Communication Bolivia*. Ministry of Sustainable Development and Planning, La Paz, Bolivia, 138 pp. Available at: http://unfccc.int

Mitchell, J.F.B. (1988) Local effects of greenhouse gases. *Nature* 332, 399–400.

Mott MacDonald (1991) Urban II Project. *Drainage Master Plan for Accra*. Ministry of Highways, Accra, Ghana.

NCCSAP (2000) *Climate Change Country Studies*. CD-ROM WL. Delft Hydraulics Z2867.

Nicholls, R. (1991) *Land Loss Scenarios for the Nile Delta, Egypt, for the Years 2060 and 2100*. A report prepared for the United States Environmental Protection Agency. Office of Policy Analysis, Washington, DC.

Nicholls, R.J. (1995) Synthesis of vulnerability analysis. In *Proceedings of World Coast 1993*. Ministry of Transport, Public Works and Water Management, The Netherlands, pp. 181–216.

Nicholls, R.J., Leatherman, S.P., Dennis, K.C. and Volonte, C.R. (1995) Impacts and responses to sea-level rise: qualitative and quantitative assessments. *Journal of Coastal Research, Fort Lauderdale* (Special Issue) 14, 26–43.

O'Brien, K. (2000) *Developing Strategies for Climate Change: the UNEP Country Case Studies on Climate Change Impacts and Adaptations Assessments*. Cicero, Oslo.

Olsen, S.B. (2003) Frameworks and indicators for assessing progress in integrated coastal management initiatives. *Ocean and Coastal Management* 46, 347–361.

Oudot, C. and Roy, C. (1991) Les sels nutritifs au voisinage de Dakar: cycle annuel et variabilité interannuelle. In: Cury, C. and Roy, C. (eds) *Pêcheries ouest-africaines. Variabilité, instabilité et changement*. Editon de l'ORSTOM (currently: Institut de Recherche pour le Développement – IRD), Paris, pp. 80–89.

Palacios, P. (1989) Determinación del Prisma de Marea y Tiempo de Renovación de los Sistemas: Río Guayas. Canal de Jambelí y Estero Salado – Canal del Morro. BA thesis, 156 pp.

Point Carbon (2004) *Carbon Market News*. Available at: http://www.pointcarbon.com

Ponzi, D. (2002) *Technical Assistance (Financed by the Government of Canada) for the Climate Change Adaptation Program for the Pacific*. Asian Development Bank, Manila, Philippines. Available at: http://adb.org

Raynal, A. (1963) *Flore et végétation des environs de Kayar (Sénégal) de la côte au lac Tanma*. Annales de la Faculté des Sciences de Dakar, Dakar, Senegal, 213 pp.

República de Costa Rica (2000) *Primera Comunicación Nacional ante la Convención Marco de Las Naciones Unidas Sobre Cambio Climático*. República de Costa Rica, San José, Costa Rica, 177 pp. Available from: http://unfccc.int

République du Sénégal (1997) *Communication Initiale du Sénégal a la Convention-Cadre des Nations-Unies Sur les Changements Climatiques (CCNUCC)*. Ministère de l'Environnement, Dakar, République du Sénégal, 118 pp. Available from http://unfccc.int

République du Sénégal (1999) *Stratégie Nationale de Mise en Œuvre (SNMO) de la Convention Cadre des Nations Unies sur les Changements Climatiques (CCNUCC)*. Ministère de l'Environnement, Dakar, République du Sénégal, 53 pp. Available from: http://unfccc.int

Sathaye, J. and Meyers, S. (eds) (1995) *Greenhouse Gas Mitigation Assessment: a Guidebook*. Kluwer Academic Publishing, Dordrecht, The Netherlands, 190 pp.

Sestini, G. (1992) Implications of climate change for the Nile Delta. In: Jeftic, I., Milliman, J.D. and Sestini, G. (eds) *Climate Change and the Mediterranean*. United Nations Development Programme (UNDP), Edward Arnold, London, pp. 535–601.

Sheppard, E., Couclelis, H., Graham, S., Harrington, J.W. and Onsrud, H. (1999) Geographies of the information society. *International Journal of Geographical Information Science* 13(8), 797–823.

Smith, M. (1992) CROPWAT *– a computer Program for Irrigation, Planning and Management*. FAO Irrigation and Drainage, Paper number 46. FAO, Rome, Italy, 126 pp.

Smith, J.B. and Lazo, J.K. (2001) A summary of climate change impact assessments from the US Country Studies Program. *Climatic Change* 50, 1–29.

Stanley, D.J. and Warner, A.G. (1993) Nile Delta: recent geological evolution and human impact. *Science* 260, 628–634.

Trochain, J.L. (1940) Contribution à l'étude de la végétation du Sénégal. *Mémoires de l'Institut Fondamentale d'Afrique Noire (IFAN)*, Dakar, 1(2), 433 pp.

Tsyban, A., Everett, J.T. and Titus, J.G. (1990) World oceans and coastal zones. In: Tegart, McG., Sheldon, G.W. and Griffiths, D.C. (eds) *Climate Change: the IPCC Impacts Assessment*. Australian Government Publishing Service, Canberra, Australia, 6.1–6.28.

UN (1992) *United Nations Framework Convention on Climate Change.* United Nations, Bonn, 33 pp. Available from: http://unfccc.int

UN (2002) *Johannesburg Summit 2002. United States of America Country Profile.* United Nations Department of Economic and Social Affairs, Division for Sustainable Development, 156 pp. Available at: http://www.un.org/esa/agenda21/natlinfo/wssd/usa.pdf

UNDP (2004) *User's Guidebook for the Adaptation Policy Framework, UNDP.* United Nations Development Programme (UNDP) – National Communications Support Unit, New York, 40 pp. Also available from: http://www.undp.org/cc/apf.htm

UNFCCC (2004a) *Reporting on Climate Change – User Manual for the Guidelines on National Communications from Non-Annex I Parties.* United Nations Framework Convention on Climate Change (UNFCCC), Bonn, 38 pp. Available from: http://unfccc.int

UNFCCC (2004b) *Compendium on Methods and Tools to Evaluate Impacts of Vulnerability and Adaptation to Climate Change.* United Nations Framework Convention on Climate Change (UNFCCC), Bonn, Germany. Available from: UNFCCC website http://unfccc.int

UNITAR (2005) *Climate Change Programme.* United Nations Institute for Training and Research (UNITAR), New York. Available at: http://www.unitar.org/ccp

USAID (2005) *Global Climate Change Program: Overview.* US Agency for International Development (USAID), Washington, DC. Available from: http://www.usaid.gov

US CSP (1999) *Climate Change: Mitigation, Vulnerability, and Adaptation in Developing Countries.* US Country Studies Program (CSP), Washington, DC.

Van Drunen, M.A. and Dorland, C. (2000) *Netherlands Climate Change Studies Assistance Programme: Approach and Management.* Institute for Environmental Studies (IVM), Vrije Universiteit, Amsterdam, 42 pp.

Verheyen, R. (2003) Adaptation funding – legal and institutional issues. In: Smith, J.B., Klein, R.J.T. and Huq, S. (eds) *Climate Change, Adaptive Capacity and Development.* Imperial College Press, London, pp. 191–216.

VROM (2001) *Third Netherlands' National Communication on Climate Change Policies.* Ministry of Housing, Spatial Planning and the Environment (VROM), The Hague, The Netherlands, 125 pp. Available from: http://unfccc.int

Weiner, D., Harris, T.M. and Craig, W.J. (2002) Community participation and geographic information systems. In: Craig, W.J., Harris, T.M. and Weiner, D. (eds) *Community Participation and Geographic Information Systems.* Taylor and Francis, New York, pp. 1–18.

Wigley, T. (2004) MAGICC *and* SCENGEN. MAGICC Model for the Assessment of Greenhouse-gas Induced Climate Change, SCENGEN – a Regional Climate SCENario GENerator. University Cooperation for Atmospheric Research, Climate and Global Dynamics Division, Climate Analysis Section, Boulder, Colorado. Available at: http://www.cgd.ucar.edu/cas/wigley/magicc/

World Energy Investment Outlook (2003) *Insights.* International Energy Agency (IEA), Paris, 550 pp.

Zdzislaw, K. (1996) Water resources management. In: Watson, R.T., Zinyowera, M.C. and Moss, R.H. (eds) *Climate Change 1995: Impacts, Adaptations and Mitigation of Climate Change.* Cambridge University Press, Cambridge, UK, pp. 469–486.

Index

Adaptation 11–16
   agriculture 15
   feasibility 134–135
   forestry 15, 56
   migration 70, 71, 126, 132
   sea level rise 29–30, 39, 40, 45, 47, 60–61, 86, 103, 129–135, 154
   warning system 64, 78, 104
   water demand 54–55, 56, 78, 109, 126–127, 157
   water supply 54, 70, 109, 126
   water resources 12–14, 24, 66–72, 76, 126–127
   see also Vulnerability
Adaptation Policy Framework 158–161, 168
Adaptive capacity 14
Air pollution 122
Avalanche 58–59, 61–64
Awareness 24, 91–92, 111, 113, 121, 153–154, 155, 161–162, 164
   authorities 40, 57, 71, 79, 133
   public 27, 32, 40, 56, 57, 63, 71, 80, 87, 109, 167

Bangladesh 128–135
Barrier analysis 10, 111–113, 119–122
Bhutan 135–138
Biodiversity see Vulnerability

Bolivia 18–26, 116–119, 119–123, 123–127, 135–138, 138–140, 141–142
Bruun rule 84

Capacity building 32, 80, 90, 92, 103, 119, 124, 133, 135, 155, 157, 162, 164, 169
Carbon dioxide 116–119, 149
Clean Development Mechanism 18, 111, 122, 149–150, 153, 169
Climate variability 22, 55, 159
   see also Extreme events
CO_2 see Carbon dioxide
Coastal management 26–32, 39–42, 43, 47, 86, 88, 91, 95, 99, 103, 133–135, 154–156
Colombia 26–34, 128–135
Common Methodology see IPCC Common Methodology
Conflict 55, 71, 135, 145
Contraction and convergence 152
Costa Rica 128–135, 135–138, 138–140, 142
Cost–benefit analysis 9, 129, 159, 168
Cost-effectiveness analysis 9, 40, 160
Crop water demand models
   CRIWAR 66, 69
   CROPWAT 49, 107
Crop yield models 14–15, 135–138
   cotton 69, 137

Crop yield models *continued*
   DSSAT3 18–19, 137
   GAP 140
   maize 19–22, 69
   potato 19–22, 106–107
   soybean 19–22
   wheat 106–107

Data availability 71, 89, 93, 105, 109, 112, 114, 118, 132, 135, 157, 164
Deforestation 6, 117, 118, 140
Disaster preparedness 161
Disaster prevention 31
Drainage capacity 93, 126, 133

Ecuador 34–42, 123–127, 128–135
Education *see* Capacity building; Awareness
Egypt 42–48, 128–135
El Niño Southern Oscillation 24, 34–39
Elevation of houses 102, 104
Emission factors 118
Emission inventory *see* Greenhouse gas emission inventory
Emission trading 150, 152
Employment 37, 45
Energy sector *see* Mitigation
Erosion 100, 125
   aeolian 85
   beach 44, 45
   coastal 27, 28, 42–45, 82, 83, 85, 88–90
   dyke 99
   trampling 73
   see also Bruun rule
Evaporation 59, 74
Extreme event 18, 24, 31, 52, 60, 79, 86, 94, 97, 109, 161, 169

Flooding 45, 60, 78, 82, 88, 97, 108
Food security 14, 16, 18, 51, 55, 82, 86, 102, 144, 157

Gender 57, 108, 109, 160
General Circulation Model 25, 38, 51, 59–60, 74–76
   see also Scenario
Geographic Information System 43, 89, 90, 94, 99, 105, 139–140, 155, 156

   see also Participatory Geographic Information System
Ghana 48–58, 123–127, 128–135, 142–143
Glacier 23, 61, 81, 75, 79
Global Environment Facility 106, 111, 148, 160
Greenhouse gas emission inventory 5, 8, 88, 116–119, 140–145
Groundwater 13, 59, 60, 79, 101, 108, 125
   see also Vulnerability, water resources
Guidelines for National Greenhouse Gas Inventories 5

Holdridge life zone classification 139–140
Hydrological model 59
   BCM 74–78
   FEFLOW 83, 84
   RAINRU 66
   WATBAL 49–55, 78
   WEAP 49–55
   STREAM 155
Hydropower 13, 49, 53–54, 66, 104, 124, 149
Hyperinflation 37, 113

Impact assessment *see* Vulnerability assessment methodology
Indicators 27, 46
Indigenous people 27
Integrated Coastal Zone Management *see* Coastal management
Inundation *see* Flooding
IPCC Common Methodology 11–12, 26, 35, 40, 44, 46, 59, 82, 88, 94, 97, 128, 154, 169
IPCC Technical Guidelines for Assessing CC Impacts and Adaptation 160, 164
Irrigation 13, 20, 45, 49, 53, 78, 101, 106–108, 124, 133
   see also Vulnerability, water resources
IVM/UNEP Handbook on Methods for CC Impact Assessment and Adaptation Strategies 12–16, 83, 106, 153, 158, 160, 164

Kazakhstan 58–64, 116–119, 128–135
Kyoto Protocol 18, 106, 151

see also United Nations Framework on Climate Change

Land subsidence 43
Land use planning *see* Spatial planning
Livelihood 16, 54, 57, 159, 167
Long Range Energy Alternative (LEAP) model 8, 9, 119

Mainstreaming 56, 154, 162, 167
Mali 64–72, 123–127, 135–138
Mangrove *see* Vulnerability, mangrove
Methane 116–119, 149
Mitigation 7, 73, 110–114, 142–145, 147–153
   energy sector 7–8, 9, 119–123, 148–149
   industry 110–113
   land use change 139
   see also Long Range Energy Alternative (LEAP) model
Mongolia 73–82, , 119–123, 143–144
Mudflow 59–59, 61–64
Multicriteria analysis 27, 44, 159, 168

National Adaptation Programmes of Action 158–161, 162
National Communication 3, 18, 32, 34, 40, 47, 48, 79, 82, 87, 95, 109, 111, 138, 140–146, 158–161, 166
Netherlands Climate Change Studies Assistance Programme x, 1, 119, 147, 161–169
Nicaragua 128–135
Nutrition *see* Food security

Participatory approach 24, 25, 32, 57, 89, 133, 153–154, 159–160, 167
Participatory Geographic Information System 156, 168
Permafrost 73, 81
Poverty reduction 57, 94, 102, 161
Precipitation *see* Scenario, climate change
Public participation *see* Participatory approach

Rainfall *see* Scenario, climate change
Relocation *see* Spatial planning
Remote sensing 89, 96, 155
Risk
   capital at 45, 85, 94, 102
   people at 45, 85, 94, 102
   species at 94
River discharge 49, 60, 74–78
Runoff *see* River discharge

Saline intrusion 27, 29, 43, 83, 91, 93, 101, 133
Salinization 78, 85, 100
Satellite images *see* Remote sensing
Scenario
   climate change 12, 14–15, 34, 49, 51–52, 74–75, 59–61, 66, 74, 83–84, 90, 94, 99, 107, 123, 136, 139, 164
   socio-economic 12, 44, 83, 93, 99, 102, 119
   see also General Circulation Model
Sea level rise (SLR) 1, 26, 34, 42, 60–61, 88–96, 99
Sea surface temperature 37
Sedimentation 91, 100, 132
Senegal 82–88, 128–135, 135–138, 144
Seven Steps Method *see* IPCC Common Methodology
Spatial planning 29, 47, 90, 154
Stakeholder *see* Participatory approach
Submersion *see* Flooding
Subsidence 101
Suriname 88–96, 116–119, 123–127, 128–135

Technology assessment 113
Temperature *see* Scenario, climate change; Sea surface temperature
Training *see* Capacity building

UNEP Country Studies on CC Impacts and Adaptation Assessments 162–166
United Nations Development Programme 148
   see also Adaptation Policy Framework
United Nations Environment Programme 4, 148

United Nations Environment Programme *continued*
see also UNEP Country Studies on CC Impacts and Adaptation Assessments
United Nations Framework on Climate Change x, 1, 11, 18, 48, 88, 97, 106, 140, 158, 161
Urban planning *see* Spatial planning
US Country Studies Program 4, 58, 65, 137, 162–166

Vietnam 97–105, 128–135, 155–156
Vulnerability
  agriculture 35, 38–39, 44, 46, 48, 53, 83, 100, 106–109, 125, 135–138, 142–145, 157–158
  archaeological sites 44
  biodiversity 34, 41, 48, 101
  coastal zone 42–43, 48, 82, 84, 88–96, 128–135, 153–157
  fishery 28, 43, 70, 84, 85, 99
  forestry 138–140, 142, 158
  livestock 73, 144
  health 54, 142
  industry 44, 84
  mangroves 28, 29–30, 34, 35, 38, 83, 85, 88–91, 99, 101
  shrimp industry 34, 35, 38–39, 85, 100
  tourism 28, 44, 62, 83, 86, 99–100
  water resources 39, 48, 55, 73–81, 93, 101, 108, 123–127, 138, 157–158
wetlands 129
Vulnerability assessment methodology 158–160, 162–169
  see also Adaptation Policy Framework; IPCC Common Methodology; IPCC Technical Guidelines for Assessing CC Impacts and Adaptation; IVM/UNEP Handbook on Methods for CC Impact Assessment and Adaptation Strategies; National Adaptation Programmes of Action

Water balance model *see* Hydrological model
Water pollution 60, 77, 91, 94, 100
Water resources *see* Adaptation; Vulnerability; Hydrological model

Yemen 106–110, 119–123, 123–127, 128–135, 135–138, 144–145

Zimbabwe 110–115, 119–123